T0134929

# Data, Engineering and Applications

Rajesh Kumar Shukla · Jitendra Agrawal ·
Sanjeev Sharma · Geetam Singh Tomer
Editors

# Data, Engineering and Applications

Volume 1

 Springer

*Editors*
Rajesh Kumar Shukla
Department of Computer Science
and Engineering
Sagar Institute of Research & Technology
(SIRT)
Bhopal, Madhya Pradesh, India

Sanjeev Sharma
School of Information Technology
Rajiv Gandhi Technological University
Bhopal, Madhya Pradesh, India

Jitendra Agrawal
School of Information Technology
Rajiv Gandhi Technical University
Bhopal, Madhya Pradesh, India

Geetam Singh Tomer
THDC Institute of Hydropower
Engineering and Technology
Tehri, Uttarakhand, India

ISBN 978-981-13-6349-8          ISBN 978-981-13-6347-4    (eBook)
https://doi.org/10.1007/978-981-13-6347-4

Library of Congress Control Number: 2019931523

This Springer imprint is published by the registered company Springer Nature Singapore Pte Ltd.
The registered company address is: 152 Beach Road, #21-01/04 Gateway East, Singapore 189721, Singapore

# Contents

## Part II   On Machine Learning

# About the Editors

**Dr. Rajesh Kumar Shukla** is a Professor and Head of the Department of Computer Science and Engineering, SIRT, Bhopal, India. With more than 20 years of teaching and research experience he has authored 8 books and has published/presented more than 40 papers in international journals and conferences. Dr. Shukla received an ISTE U.P. Government National Award in 2015 and various prestigious awards from the Computer Society of India. His research interests include recommendation systems and machine learning. He is fellow of IETE, a senior member of IEEE, a life member of ISTE, ISCA, and a member of ACM and IE(I).

**Dr. Jitendra Agrawal** is a member of the faculty at the Department of Computer Science and Engineering, Rajiv Gandhi Proudyogiki Vishwavidyalaya, Bhopal, India. His research interests include data mining and computational intelligence. He has authored 2 books and published more than 60 papers in international journals and conferences. Dr. Agrawal is a senior member of IEEE, life member of CSI, ISTE and member of IAENG. He has served as a part of the program committees for several international conferences organised in countries such as the USA, India, New Zealand, Korea, Indonesia and Thailand.

**Dr. Sanjeev Sharma** is a Professor and Head of the School of Information Technology, Rajiv Gandhi Proudyogiki Vishwavidyalaya, Bhopal, MP, India. He has over 29 years of teaching and research experience and received the World Education Congress Best Teacher Award in Information Technology. His research interests include mobile computing, ad-hoc networks, image processing and information security. He has edited proceedings of several national and international conferences and published more than 150 research papers in reputed journals. He is a member of IEEE, CSI, ISTE and IAENG.

**Dr. Geetam Singh Tomer** is the Director of THDC Institute of Hydropower Engineering and Technology (Government of Uttarakhand), Tehri, India. He received the International Plato award for Educational Achievements in 2009. He completed his doctorate in Electronics Engineering from RGPV Bhopal and post-doctorate from the University of Kent, UK.

Dr. Tomar has more than 30 years of teaching and research experience and has published over 200 research papers in reputed journals, as well as 11 books and 7 book chapters. He is a senior member of IEEE, ACM and IACSIT, a fellow of IETE and IE(I), and a member of CSI and ISTE. He has also edited the proceedings of more than 20 IEEE conferences and has been the general chair of over 30 conferences.

# Part I
# On Data Mining and Social Networking

# A Review of Recommender System and Related Dimensions

Taushif Anwar and V. Uma

## 1 Introduction

Recommender system (RS) is subclass of information filtering system that deals with the problem of information overhead and helps in decision making in large information spaces [1]. In other words, we need to develop more personalized form of information access and discovery that has the capability to understand the needs of users and respond to these needs in a more efficient and objective way.

RSs are used by business organization (e-commerce) to recommend product to their users. The product can be suggested on the basis of past buying behavior, likes, comments, demographics of the customer, top seller on a site, etc. Nowadays, most of the e-commerce companies provide web recommendation to the user by systematically enabling RS at the back end.

RSs have been designed and developed using heuristics approaches [2], data mining techniques [3] and pattern mining (association rule, similarity measure) [4]. Popular RS includes Netflix and MovieLens [4, 5] for movies, Amazon.com [6] for books, CDs, and many other products, Entrée for restaurant, and Jester [7] for jokes.

### 1.1 Motivation and Problem Explanation

Nowadays, due to the hasten evolution of diverse technology, accumulation and generation of digital data can be accomplished effortlessly. Revolution in several domains like information retrieval, approximation theory, machine learning, statis-

T. Anwar (✉) · V. Uma
Department of Computer Science, Pondicherry University, Puducherry 605014, India
e-mail: taushif21589@gmail.com

V. Uma
e-mail: umabskr@gmail.com

© Springer Nature Singapore Pte Ltd. 2019
R. K. Shukla et al. (eds.), *Data, Engineering and Applications*,
https://doi.org/10.1007/978-981-13-6347-4_1

tics, and pattern recognition have given chance to unfold and mine unknown as well as interesting patterns from the data. RS is a field of information filtering, which has gained much attention in recent years.

Existence of data in various formats, viz. multimedia elements of web pages, URL logs, user likes, most viewed, or purchased item details, poses a great challenge to RS. Based on these data, filtering, a useful and interesting process removes unwanted and redundant information from the information stream and delivers the information that the user is likely to search. Filtering method is chosen based on how the data is being mined or analyzed. The three filtering methods used in RS are content-based filtering (CBF), collaborative filtering (CF), and hybrid filtering (HF).

CBF suggests item based on the individual's choice made in the past and also based on the most viewed, bought, liked, and positively ranked items. The drawback of this approach is the cold-start problem. Initially, when sufficient information has not yet been gathered, which is otherwise called as cold-start problem, the CBF system cannot be effective as it requires large amount of data/information about users/items for precise and accurate filtering.

CF approach works on retrieving and analyzing enormous volume of information on user's behaviors, activities, and recommending the user based on their information similarity with other users [1]. In CF, problem occurs when there are potentially few users, very less preferences, and many unrated items as the condition becomes sparse. This leads to data sparsity problem, and hence, it becomes very difficult to find users with similar interest.

HF approach attempts to merge different approaches for removing the data sparsity problem. Several filtering approaches can be merged in different ways. HF approach was mainly introduced to handle the problems of traditional RS. Mostly HF is applied on descriptions of items and user profiles for finding users with similar interests, followed by the use of collaborative filtering to make recommendations.

In this paper, we have emphasized on RS metrics and explained the concepts that will provide a better understanding of the system. Simultaneously, we have pointed out the dimensions associated with this system. Section 2 explains the literature reviews related to RS, and Sect. 3 explains the general concepts involved in a RS. In Sect. 4, the metrics and its importance in RS are presented. Section 5 highlights the various dimensions in RS. Section 6 concludes the paper by highlighting the various issues in RS.

## 2 Literature Review

The foundation of the RSs started with research papers on CF by Resnick et al. [8] in 1994 and Hill et al. [9] in 1995. Several methods are used in RSs for clustering, regression, and classification. RS has dimensions such as novelty, serendipity, correctness, diversity, privacy, usability, and user preference which are useful in providing proper recommendations.

Bobadilla et al. [10] presented a survey paper related to RS. This article provides an overview of RS, classification, CF and its algorithm, possible areas of the future implementation, and importance of RS. The evaluation of RS started from the first generation. In this generation, the details regarding the most liked, purchased, viewed, items, user's record, and user's item preferences are collected from the websites. The second generation mainly uses details collected from social networking websites, and third generation mainly focuses on information collected by integrating devices.

Shiyu et al. [11] presented a paper related to streaming of RS where data stream available in different properties and data gathered in continuously and rapidly. Authors have also defined mainly three challenges which are real-time updation, concept shifting, and unknown size data. For handling these issues, they have proposed sRec algorithm which can manage stream inputs through continuous-time random process and can provide real-time recommendations. Experimental testing is done using MovieLens and Netflix data sets.

He et al. [12] presented a survey paper related to RS with visualization approach which increases the accuracy, supports transparency, and user control. In this paper, different visualization approaches are suggested which can help researchers and practitioners in choosing appropriate visualization approach.

Cheng et al. [13] proposed a RS using wide and deep learning structure to combine the strengths of both models. Memorization is improved by wide linear model using cross-product features and deep neural networks which improve the generalization in RS.

# 3 Recommender System Model

RS suggests and recommends items to users on the basis of their interest and behaviors [14]. A simple model of a conventional recommender is shown in Fig. 1, and essential role of items and users in the recommendation process is highlighted.

In Fig. 1, items with its description (attributes) and users with their profiles form the basis of item and profile exploitation, respectively. Profile-item matching is performed and by filtering top-N items recommended. While content-based filtering is done in the above-mentioned manner, collaborative filtering is done when only user profiles are available. Prediction of items is done by matching user interest with similar user's (neighbors) interest.

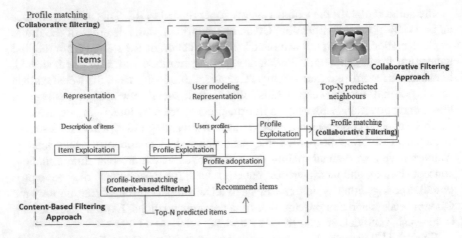

**Fig. 1** Simple model of recommender process

## 4　Evaluation Metrics for Recommendation Algorithms

### 4.1　For Predict on User Ratings

In RS approach, quality is measured by analyzing the recommendations made for a test set containing familiar user ratings. This is done by calculating the mean absolute error (MAE) [15, 16] which reflects the predictive accuracy [17] where comparison is done between predicted and actual user rating.

MAE calculates the average absolute deviancy of expected value from user value, viz. average absolute error.

$$MAE = \frac{\Sigma_{(x,y)}\left(P_{x,y} - A_{x,y}\right)}{N} \tag{1}$$

where $P_{x,y}$ is predicted value of the user $x$ for the item $y$, $A_{x,y}$ is the actual (real) value of user $x$ for item $y$, and $N$ is the total number of rating in the test set.

Root-mean-square error (RMSE) is widely accepted method for scoring an algorithm. It calculates larger absolute error.

$$RMSE = \sqrt{\frac{\Sigma_{(x,y)}\left(P_{x,y} - A_{x,y}\right)^2}{N}} \tag{2}$$

To avoid the scale dependency, normalized MAE and normalized RMSE have been proposed. NMAE is the ratio of MAE and the diversity of the maximum and minimum values of rating.

$$\text{Normalized MAE} = \frac{\text{MAE}}{r_{\max} - r_{\min}} \tag{3}$$

NRMSE is the ratio of RMSE and the diversity of the maximum and minimum values of rating.

$$\text{Normalized RMSE} = \frac{\text{RMSE}}{r_{\max} - r_{\min}} \tag{4}$$

Predictive accuracy metrics behave uniformly for all items. So, mostly RS approaches are mainly related to suggest the items that a user will like.

For example, consider a problem of ranking Python files for recommendation based on the code searched. Four files are suggested to a user with predicted value (rating) 5, 5, 3, and 4 in a 1–5 (Likert) scale rating system, but users actual rating are 5, 3, 4, and 3. Hence, the error measures can be computed as follows:

$$\text{MAE} = \frac{((5-5) + (5-3) + (3-4) + (4-3))}{4} = 0.5 \tag{5}$$

$$\text{RMSE} = \sqrt{\frac{((5-5)^2 + (5-3)^2 + (3-4)^2 + (4-3)^2)}{4}} = \sqrt{\frac{6}{4}} \approx 1.224 \tag{6}$$

$$\text{Normalized MAE} = \frac{\text{MAE}}{5-1} = \frac{0.50}{4} = 0.125 \tag{7}$$

$$\text{Normalized RMSE} = \frac{\text{RMAE}}{5-1} = \frac{1.224}{4} \approx 0.306 \tag{8}$$

RMSE has the benefit of reducing large errors, and here, in this example value of MAE is 0.5 and value of RMSE is 1.224. In this case, error in RMSE is more than MAE. RMSE can be greater in some cases, and in some cases, it may be equal to MAE.

## 5 Dimensions of Recommender System

The dimensions of RS are listed in (Table 1).

**Novelty**: Novelty is highly desirable and important feature of RS. In this recommendation approach, the user does not have a clear idea about items. In the context of a single metric encompassing, novelty and accuracy both should be useful for a more conclusive and comprehensive comparison of recommendation list. It is also correlated with the emotional replay of users toward suggestions, and as an outcome calculating this dimension is a challenge.

**Serendipity**: RS accidently suggests novel items which are unexpected but useful. There is different type of emotional reply linked to serendipity which is very challenging to manipulate using any of the metrics. The idea of serendipity is that

**Table 1** Dimensions: definition and categorization

|   | Dimension | Functionality | Type |
|---|-----------|---------------|------|
| 1 | Novelty | Recommendation approaches suggest items which are unknown or new to users | User-centric |
| 2 | Serendipity | To what extent does the system accidently suggest items unexpected but useful? | User-centric |
| 3 | Correctness | Calculates truthfulness of a system | Recommendation-centric |
| 4 | Diversity | How unrelated are the suggested items in a list? | Recommendation-centric |
| 5 | Stability | How reliable are the suggestions over an amount of time? | System-centric |
| 6 | Privacy | Are there any hazards to user secrecy or privacy? | System-centric |
| 7 | Usability | How functional is the recommendation approach? | Delivery-centric |
| 8 | User preference | User perception about the recommendation approach | Delivery-centric |

user has suggested some items which do not appear to be connected and that will subsequently become interesting [18]. The main problem with serendipity in recommender approach is that the quality of recommendation deteriorates, and so it is very necessary to find right balance between novelty and similarity.

**Correctness**: Success of RS approach depends on its correctness, and hence, it plays a key role in the recommendation. Correctness calculates how truthful a system is by evaluating the degree to which how close the measured values are to each other (precision) as well as the number of positive or relevant items suggested (recall).

**Diversity**: Diversity represents the correlation between the recommended items. In a generic method, the diversity among list of items can be measured, i.e., in terms of the objective difference (e.g., as pair-wise dissimilarity/unlikeness).

**Stability**: In RS, stability plays very imperative role, and it is described as the capability of a system to be consistent or unfailing in predictions made on similar items using similar algorithms so that new rating is in complete association with system's earlier approximations. Generally, stability measures the level of core consistency and reliability among predictions made by the given recommendation algorithm. In RS, inconsistency may have a bad impression on user's trust.

**Privacy**: Nowadays, privacy is the main concern which is rarely considered in RS. RS suggests an item on the basis of user's behavioral data which is collected from different sources such as explicit product rating, viewed, comments, and implicit purchase history. Privacy directly affects two desirable qualities of recommender approaches: their scalability and their predictive accuracy. In most of the methods, one thing is common: RS collects information related to the users and stores it in a centralized storage.

**Usability**: In RS, our aim is to provide well-organized and effective recommendation which delivers some degree of satisfaction to the end user. Generally, RS interacts with user interface, and so content represented in user interface plays a vital role in acceptance of recommendation [19]. Interface really impacts the overall usability of RS.

**User preference**: Nowadays, we are measuring performance of RS in various ways. Preferences are the way by which our choices can be monitored, and the items preferred can be discriminated from those that are not preferred. The minimum level of assessment in the recommendation approach is based on the awareness of the individual about that system.

## 6 Conclusion

Much research has been done on RSs over the last few decades. Content-based filtering, collaborative filtering, and hybrid filtering approaches are used in recommendation systems. The existing approaches consider handling certain dimensions and hence do not provide proper recommendations. This demands extensive work, and hence, providing a recommender approach with higher accuracy, precision, and recall remains a challenge.

## References

1. Konstan, J.A., Riedl, J.: Recommender systems from algorithms to user experience. User Model User-Adapt Interact **22**, 101–123. Springer Science + Business Media (2012)
2. Joaquin, D., Naohiro, I.: Memory-based weighted-majority prediction for recommender systems. In: 1999 SIGIR Workshop on Recommender Systems, pp. 1–5. University of California, Berkeley (1999)
3. Hofmann, T.: Latent semantic models for collaborative filtering. ACM Trans. Inf. Syst. **22**, 89–115 (2004)
4. Huang, C.L., Huang, W.L.: Handling sequential pattern decay: developing a two-stage collaborative recommender system. Electron. Commer. Res. Appl. **8**, 117–129 (2009)
5. Miller, B.N., Albert, I., Lam, S.K., Konstan, J.A., Riedl, J.: MovieLens unplugged: experiences with an occasionally connected recommender system. In: Proceedings of the 8th International Conference on Intelligent User Interfaces, Miami, Florida, USA (2003)
6. Linden, G., Smith, B., York, J.: Amazon.com recommendations: item-to-item collaborative filtering. IEEE Internet Comput. **7**, 76–80 (2003)
7. Goldberg, K., Roeder, T., Gupta, D., Perkins, C.: Eigentaste: a constant time collaborative filtering algorithm. Inf. Retrieval J. **4**, 133–151 (2001)
8. Resnick, P., Iakovou, N., Sushak, M., Bergstrom, P., Riedl, J.: GroupLens: an open architecture for collaborative filtering of netnews. In: Proceeding of the 1994 ACM Conference on Computer Supported Cooperative Work, pp. 175–186 (1994)

9. Hill, W., Stead, L., Rosenstein, M., Furnas, G.: Recommending and evaluating choices in a virtual community of use. In: The Proceedings of the SIGCHI Conference on Human Factors in Computing Systems, pp. 194–201 (1995)
10. Bobadilla, J., Ortega, F., Hernando, A., GutiéRrez, A.: Recommender systems survey. Knowl.-Based Syst. 109–132 (2013)
11. Shiyu, C., Yang Zhang, Z., Jiliang, T.: Streaming recommender systems. In: Proceedings of the 26th International Conference on World Wide Web ACM, pp. 381–389 (2017)
12. He, C., Parra, D., Verbert, K.: Interactive recommender systems: a survey of the state of the art and future research challenges and opportunities. Expert Syst. Appl. 9–27 (2016)
13. Cheng, H.-T., Koc, L., Harmsen, J., Shaked, T., Chandra, T., Aradhye, H., Anderson, G., Corrado, G., Chai, W., Ispir, M., et al.: Wide & deep learning for recommender systems. In: Proceeding of the 1st Workshop on Deep Learning for Recommender Systems, pp. 7–10 (2016)
14. Pu, P., Chen, L., Hu, R.: Evaluating recommender systems from the user's perspective: survey of the state of the art. User Model. User-Adap. Inter. 22, 317–355 (2012)
15. Koren, Y., Bell, R., Volinsky, C.: Matrix factorization techniques for recommender systems. Computer 42, 30–37 (2009)
16. Su, X., Khoshgoftaar, T.M.: A survey of collaborative filtering techniques. Adv. Artif. Intell. 1–19 (2009)
17. Herlocker, J.L., Konstan, J.A., Terveen, L.G., Riedl, J.T.: Evaluating collaborative filtering recommender systems. ACM Trans. Inf. Syst. 22, 5–53 (2004)
18. Iaquinta, L., Gemmis, M.D., Lops, P., Semeraro, G., Filannino, M., Molino, P.: Introducing serendipity in a content-based recommender system. In: Proceedings of the 2008 8th International Conference on Hybrid Intelligent Systems, pp. 168–173 (2008)
19. Ozok, A.A., Fan, Q., Norcio, A.F.: Design guidelines for effective recommender system interfaces based on a usability criteria conceptual model: results from a college student population. Behav. Inf. Technol. 29, 57–83 (2010)

# Collaborative Filtering Techniques in Recommendation Systems

Sandeep K. Raghuwanshi and R. K. Pateriya

## 1 Introduction

Recommendation systems are information filtering systems that urge to predict preferences that user might have for an item over other. Recommendation systems are very popular in applications like movies, books, research articles, search queries, social tags, product, financial services, restaurants, twitter pages, job, university, friends and what not. To increase product sales is the primary goal of recommendation system by bringing a relevant item to the user and thus increasing the overall profit, which covers the functional goal of recommendation system such as [1]—relevancy, serendipity and diversity. Most popular recommender systems of today are Group Lens recommender system, Amazon.com recommender system, Netflix Movie recommender system, Google News personalisation system, Facebook friend recommendations, link prediction recommender system [1].

First recommendation system was developed in 1992 by Goldberg, Nichols, Oki and Terry. This was called Tapestry which allows users to rate an item good or bad and further used keyword filtering for recommendation [2–4]. Thus, recommendation system works on available information in any form and then applies different filtering techniques to find the most appropriate choice (like the movie, show, web page, scientific literature and news that a user might have interest in). The recommendation system makes use of data mining techniques [4, 5] and prediction algorithm to find out user's interest in information, item and their other interests. Later on, several recommendation systems developed which use different filterings to lure their customers and make them feel more attended (Fig. 1).

S. K. Raghuwanshi (✉) · R. K. Pateriya
Computer Science and Engineering, Maulana Azad National Institute of Technology,
Bhopal, M. P, India
e-mail: sraghwuanshi@gmail.com

R. K. Pateriya
e-mail: pateriyark@gmail.com

© Springer Nature Singapore Pte Ltd. 2019
R. K. Shukla et al. (eds.), *Data, Engineering and Applications*,
https://doi.org/10.1007/978-981-13-6347-4_2

**Fig. 1** Recommendations
and recommender system

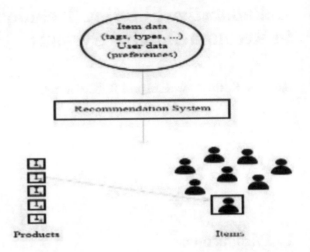

The reason for many companies care about recommendation system is to deliver actual value to their customer. Recommender systems provide a scalable way of personalising content for users in scenarios with many items. It engages many scientists, since it is a major problem of data science, a perfect intersection of software engineering, machine learning and statistics. Recommender systems are an effective tool for personalisation. Since it is based on actual user behaviour, users can make decisions directly based on the results. These systems work on unstructured and dynamically changing data because of which predictions are more specific and up to date.

Although recommender systems are application-specific and require specific filtering process, few properties must be addressed by all of them [6] like user preference, prediction accuracy, confidence score, user's trust on a recommendation system.

The rest of this paper is organised as follows: first section deals with the introduction of recommender system with their applicability and importance in present era. Section 2 presents the goals and critical challenges of recommendation systems. Section 3 presents the classification of recommendation system based on the approach to build recommendation engine. This section presents a brief introduction of content-based recommender system with collaborative techniques in detail and presents two different approaches of collaborative as memory-based and model-based systems. Section 4 presents experimental set-up to show the methods implementation and results. Section 5 gives the conclusion of work and possible future scope.

## 2 Goals and Critical Challenges

### 2.1 Goals

Recommender systems are used in different fields, from e-commerce to government applications. Most widely used application of recommendation system comes from e-commerce where companies are competing for enhancing their sales and improve user experience. By recommending interested and preferred items to users' recommender system helps merchants to increase their profit. Apart from this, the general operational and technical goals of recommendation systems are as follows:

**Relevance**: The most common operational goal of recommender system is to provide or recommend relevant items to the users. Users are more likely to purchase or opt in items which are of his/her preferences.

**Novelty**: Recommendation systems are supposed to provide novel or new items each time. The system should not repeatedly show popular items as this may also leads to reduction in user interest [7].

**Serendipity**: Serendipity is notion to define somewhat unexpected recommendation. It is different from novelty as it is truly surprising to user instead that they did not know about before [8].

**Diversity**: Recommendation system generally recommends list of similar items which increases the chance that user might not like any time. So, diversity is one of the important goals of recommender system which supports range of items for recommendation.

### 2.2 Challenges

Following are the critical challenges of recommendation systems:

**Scalability**: Most collaborative filtering techniques show poor performance with an increase in user and item base.

**Grey Sheep**: Grey sheep denotes the group of peoples whose opinions do not match with any group of people. These users basically create a problem in the smooth functioning of recommendation system [9].

**Synonymy**: Most recommender systems face problem to predict accurately the items which are same in features but have different names [10, 11].

**Cold Start**: New users and items suffer from accurate prediction as not much information is available to start the system [12].

**Privacy Breach**: Privacy has always been the biggest challenge of a recommender system. While providing an accurate prediction of user system demand to get personalised information of the user.

**Shilling Attack**: Recommendation is a public activity, so people get biased for their feedbacks and give millions of positive reviews for their own products or items and sometimes negative views of their competitors [13].

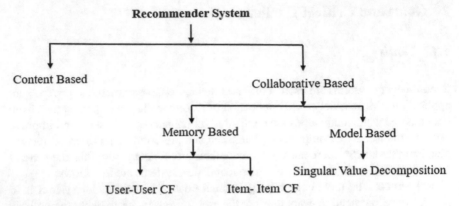

**Fig. 2** Classification of recommendation system

## 3    Classification

Depending on the type of input used to make recommendations, recommender systems are classified into several categories. Out of which, most commonly used techniques are content-based filtering and collaborative filtering. This paper accentuates various recommendation systems used today with their pitfalls and comparative analysis of two major recommendation models (NN and Latent factor model) (Fig. 2).

### 3.1    Content-Based Filtering System

Content-based filtering is the most common type of filtering system used. These systems work on rating, which a user gave while creating a profile to get initial information about a user in order to avoid not knowing a new user [14]. To create user profile, two types of information are mainly focused: user's preferences and users interaction with recommendation system. It simply recommends items on the basis of comparison between the content of the item and a user's profile. Engines in these systems compare positively rated item by a user with the item he/she did not rate yet. The items with maximum similarities will then be recommended to the users. Different distances are used for measuring distances/similarities between user's choice and among items in the database (Fig. 3).

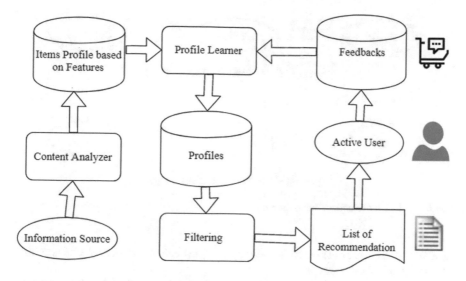

**Fig. 3** Content-based filtering workflow

## 3.2 Collaborative Filtering

Collaborative filtering algorithm works by collecting and analysing a large amount of information on user behaviour, their preferences and their activities. Collaborative filtering is capable of recommending complex items more accurately because it does not reside on content analysed by machine. Such recommendation systems work on assumption that a user agreed in past will be interested in future as well and they are more probable to like similar kind of item.

Collaborative filtering techniques use distance feature to calculate similarities between user's choice and among items in database such as cosine distance, Pearson distance and Euclidean distance. We have implemented cosine and Pearson similarities to calculate similarity. Cosine similarity— this uses a coordinate space in which items are represented as a vector. It measures the angle between vectors and gives out their cosine values [5]. Pearson distance—it is a measure of linear correlation between two variables.

$$S_{pearson} = \frac{\sum_{i=1}^{n}(r_{u,i} - \hat{r}_u) \times (r_{v,i} - \hat{r}_v)}{\sqrt{\sum_{i=1}^{n}(r_{u,i} - \hat{r}_u)^2 \times (r_{v,i} - \hat{r}_v)^2}} \tag{1}$$

$$S_{cosine} = \frac{\sum_{i=1}^{n} r_{u,i} \cdot r_{v,i}}{\sqrt{\sum_{i=1}^{n}(r_{u,i})^2 \times (r_{v,i})^2}} \tag{2}$$

The idea is to create a community that shares a common interest [15]. Such users form a neighbourhood. And thus, a user gets a recommendation for items that he/she

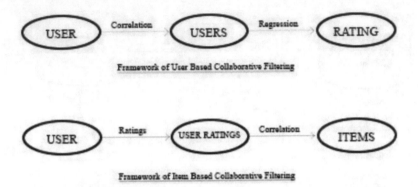

Fig. 4  Framework of collaborative filtering

have not rated before but rated positively by users in his/her community. Collaborative filtering is of following types.

Memory-based approach: they are also called as neighbourhood-based collaborative filtering algorithms, in which the ratings of user–item combinations are predicted on the basis of their neighbourhoods which include user–user-based collaborative filtering and item–item-based collaborative filtering [16–18].

*User-based Collaborative Filtering*: In this, the rating predictions are calculated based on similar minded users of the target user. To predict rating preference for user *A*, the idea is to find top *k* similar users of *A* and compute weighted average of ratings of peer group.

*Item-based Collaborative Filtering*: In this, item similarity is used to determine rating prediction for target user. The idea is to find a set of similar items for which prediction was sought and then use these items' rating to compute final prediction of user to item.

Model-based approach: In this, machine learning and data mining methods are used in the context of predictive models. For example: decision tree, rule-based model, Bayesian model and latent factor model. In this paper, we have implemented latent factor model, using SVD (singular value decomposition) (Fig. 4).

## 4   Experimental Set-up and Results

This paper illustrates the implementation of two basic recommender systems (memory-based and model-based) and compares their performance on the basis of various evaluation parameters.

## 4.1 Data set

This paper has used the following data sets to implement recommendation algorithms.

**Movie lens**: This data set describes 5-star rating and free-text tagging activity from movie lens, a movie recommendation service. It contains 100,004 ratings and 1296 tag applications across 9125 movies. These data were created by 671 users between 09 January 1995 and 16 October 2016. This data set was generated on 17 October 2016 [19].

**Jester**: Over 4.1 million continuous ratings ($-10.00$ to $+10.00$) of 100 jokes from 73,421 users were collected between April 1999 and May 2003 [20].

## 4.2 Working Process

### 4.2.1 Memory-Based Collaborative Filtering: User-Based Collaborative Filtering

User-based collaborative filtering is based on the assumption that similar users with similar preferences will rate their choices similarly. One has to find that similarity and predict missing ratings for that user. When missing, ratings are known, and we can also recommend user items as per his/her taste [21].

Item-based collaborative filtering—This looks into the sets of items that target user has rated and compute how similar they are to the target item $i$ and then select more similar item $k$, and also compares their corresponding similarities. The prediction is then computed by taking a weighted average of the target user's ratings on these similar items [22].

$$P_{u,j} = \hat{r}_u + K \sum_{i=1}^{n} S(u,i) \times \left(r_{i,j} - \hat{r}_i\right) \qquad (3)$$

*Model-based filtering*: SVD—Singular value decomposition is a well-established technique for identifying latent semantic factors in information retrieval. Collaborative filtering uses SVD by factoring user item rating matrix.

Let the user item rating matrix is described as $R_{n*m}$ with $N$ number of users' rate $M$ items, and *Rij* describes the rating of item $j$ given by user $i$. For a matrix $R$, its SVD is factorisation of $R$ into three matrices such that·

$$R = P \Sigma Q^{\mathrm{T}} \qquad (4)$$

where $\sum$ is the diagonal matrix whose values $\sigma i$ are the singular values of decomposition, and both $P$ and $Q$ are the orthogonal matrices, which means $P^{\mathrm{T}}P = I_{nxn}$ and $Q^{\mathrm{T}}Q = I_{mxm}$. Originally, matrix $P$ is $n \times k$, $\sum$ is $k \times k$, and $Q$ is $m \times k$, where $R$ is $n \times m$ and has rank $k$.

The SVD represents an expansion of the original rating matrix in a coordinate system where the covariance matrix is diagonal. Matrix $P$ represents user latent values, and matrix $Q$ gives the item latent feature for given rating matrix [23].

## 5   Results

This paper considered the following parameters for evaluation and for a comparison of different algorithms against a data set, which has been shown in Table 1 (Fig. 5).

### 5.1   RMSE

The RMSE (root-mean-square error) is computed by averaging the square of the differences between UV and the utility matrix, in those elements where the utility matrix is nonblank. The square root of this average is the RMSE [24].

$$\text{RMSE} = \sqrt{\frac{1}{N}\sum_{i=1}^{n}(P_i - R_i)^2} \qquad (5)$$

**Table 1**  Results

| Technique | Data set | RMSE | MAE | F-measure |
|-----------|----------|------|-----|-----------|
| UBCF | Movie lens | 0.504 | 0.253 | 0.35179 |
| | Jester | 4.238 | 3.336 | 2.42 |
| | Movie lens | 0.483 | 0.206 | 0.3056 |
| IBCF | Jester | 4.241 | 3.335 | 1.860 |
| | Movie lens | 0.560 | 0.225 | 0.349 |
| SVD | Jester | 0.949 | 3.507 | 1.772 |

**Fig. 5**  Comparison of RMSE value of UBCF, IBCF and SVD on movie lens data set and Jester data set

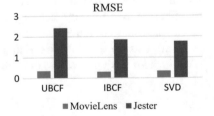

## 5.2 MAE

The mean absolute error is an average of the absolute errors (Fig. 6).

$$\text{MAE} = \frac{1}{N} \sum_{i=1}^{n} |P_i - R_i| \tag{6}$$

## 5.3 F-Measure

Metric combines Precision and Recall into a single value for comparison pulse.

$$F\text{-Measure} = \frac{2 * \text{Precision} * \text{Recall}}{(\text{Precision} + \text{Recall})} \tag{7}$$

Precision is the measure of exactness. It determines the fraction of relevant items retrieved out of all items.

$$\text{Precision} = \frac{\text{True Positive}}{(\text{True Positive} + \text{False Positive})} \tag{8}$$

Recall is the measure of completeness. It determines the fraction of relevant items retrieved out of all items (Fig. 7).

$$\text{Recall} = \frac{\text{True Positive}}{(\text{True Positive} + \text{False Negative})} \tag{9}$$

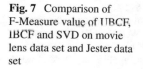

**Fig. 6** Comparison of MAE value of UBCF, IBCF and SVD on movie lens data set and Jester data set

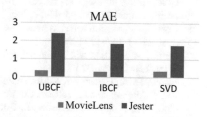

**Fig. 7** Comparison of F-Measure value of UBCF, IBCF and SVD on movie lens data set and Jester data set

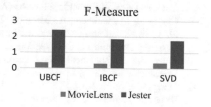

# 6 Conclusion and Future Scope

Recommendation system serves as a useful tool for users in expanding their interest and their experience over the Internet. Recommendation accelerates profits for developer and business person by knowing their customers well serving them best. Along with mobiles and computers, they open new security doors for the automobile industry and devices used on daily basis. Among several solutions and facilities, there are some issues related to available recommendation system that needs to be addressed specifically to take most out of them [25].

The recommendation can be made more complete and accurate by using latest data mining techniques and machine learning approach. Incorporating artificial intelligence into underlying algorithm strengthens the system to a greater extent as it helps in knowing the audience well and enough. Further improvisation is required so that recommendation system can do the intended job without compromising privacy and information leakage as mentioned above. All these factors imply that we are still in the urge to make promising systems, and there is way more to go for their development [26].

# References

1. Aggarwal, C.C.: Recommender Systems. Springer International Publishing, Switzerland (2016). https://doi.org/10.1007/978-3-319-29659-3
2. Paul, R., Neophytos, I., Mitesh, S., Bergstrom, P., Riedl, J.: GroupLens: an open architecture for collaborative filtering of netnews. In: Proceedings of the 1994 ACM Conference on Computer Supported Cooperative Work Chapel Hill, North Carolina, United States, pp. 175–186 (1994)
3. Witten, I.H., Frank, I.: Data Mining. Morgan Kaufman Publishers, San Francisco (2000)
4. Jhon Breese, S., Heckerman, D., Kadie, C.: Empirical analysis of predictive algorithms for collaborative filtering. In: Proceedings of the Fourteenth Annual Conference on Uncertainty in Artificial Intelligence, pp. 43–52, July (1998)
5. Deshpande, M., Karypis, G.: Item-based top-N recommendation algorithms. ACM Trans. Inf. Syst. 22(1), 143–177 (2004)
6. Aamir, M., Bhusry, M.: Recommendation system: state of the art approach. Int. J. Comput. Appl. 120(12), 25–32 (2015)
7. Fleder, D.M., Hosanagar, K.: Recommender systems and their impact on sales diversity. In: ACM Conference on Electronic Commerce, pp. 192–199 (2007)
8. Good, N., Schafer, J., Konstan, J., Borchers, A., Sarwar, B., Herlocker, J., Riedl, J.: Combining collaborative filtering with personal agents for better recommendations. In: National Conference on Artificial Intelligence (AAAI/IAAI), pp. 439–446 (1999)
9. Claypool, M., Gokhale, A., Miranda, T., et al.: Combining content-based and collaborative filters in an online newspaper. In: Proceedings of the SIGIR Workshop on Recommender Systems: algorithms and Evaluation, Berkeley, Calif, USA (1999)
10. Jones, S.K.: A statistical interpretation of term specificity and its applications in retrieval. J. Documentation 28(1), 11–21 (1972)
11. Gong, M., Xu, Z., Xu, L., Li, Y., Chen, L.: Recommending web service based on user relationships and preferences. In: 20th International Conference on Web Services, IEEE (2013)
12. Rana, M.C.: Survey paper on recommendation system. Int. J. Comput. Sci. Inf. Technol. 3(2), 3460–3462 (2012)

13. Resnick, P., Varian, H.R.: Recommender systems. Commun. ACM **40**(3), 56–58 (1997)
14. Balabanovi, M., Shoham, Y.: Fab: content based, collaborative recommendation. Mag. Comm. ACM **40**(3), 66–72 (1997)
15. Puntheeranurak, S., Chaiwitooanukool, T.: An item-based collaborative filtering method using item-based hybrid similarity. In: Proceedings of the IEEE 2nd International Conference on Software Engineering and Service Science (ICSESS), pp. 469–472. ISBN: 978-1-4244-9699-0 (2011)
16. Miyahara, K., Pazzani, M.J.: Collaborative filtering with the simple Bayesian classifier. In: Pacific Rim International Conference on Artificial Intelligence, pp. 679–689 (2000)
17. Ghani, R., Fano, A.: Building recommender systems using a knowledge base of product semantics. In: 2nd International Conference on Adaptive Hypermedia and Adaptive Web-Based Systems, pp. 27–29 (2002)
18. Elgohary, A., Nomir, H., Sabek, I., Samir, M., Badawy, M., Yousri, N.A.: Wiki-rec: A semantic-based recommendation system using wikipedia as an ontology. Intell. Syst. Des. Appl. (ISDA) (2010)
19. http://files.grouplens.org/datasets/movielens/ml-latest-small-README.html
20. http://eigentaste.berkeley.edu/dataset/
21. Sarwar, B.M., Karypis, G., Konstan, J.A., Riedl, J.: Analysis of recommendation algorithms for E-commerce. In: Proceedings of 2nd ACM Conference on Electronic Commerce Minnesota USA, pp. 158–167 (2000)
22. Karypis, G.: Evaluation of item-based top-N recommendation algorithms. In: Proceedings of the International Conference on Information and Knowledge Management (CIKM '01), Atlanta, GA, USA, pp. 247–254 (2001)
23. Koren, Y.: Collaborative filtering with temporal dynamics. Commun. ACM **53**(4), 89–97 (2010)
24. http://infolab.stanford.edu/~ullman/mmds/ch9.pdf
25. Pronk, V., Verhaegh, W., Proidl, A., Tiemann, M.: Incorporating user control into recommender systems based on naive bayesian classification. In: RecSys'07: Proceedings of the 2007 ACM Conference on Recommender Systems, pp. 73–80 (2007)
26. Kanawati, R., Karoui, H.: A p 2p collaborative bibliography recommender system. In: Proceedings Fourth International Conference on Internet and Web Applications and Services, Washington DC, USA. IEEE Comput. Soc. 90–96 (2009)

# Predicting Users' Interest Through ELM-Based Collaborative Filtering

Shweta Tyagi, Pratibha Yadav, Moni Arora and Pooja Vashisth

## 1 Introduction

Recommender systems (RSs) have arisen as a significant tool, which aim at assisting the users to find desired and appropriate recommendations based on users' preferences [1, 2]. Since its emergence in the mid-1990s, an extensive amount of research has been taken place with practical applications in the domain of movies, jokes, music, books, Web sites, restaurants, e-learning material, electronic products, digital products, and many more [3–5]. Still, it is a problem-rich research area and getting more attention of researchers, academia, and industry.

In the literature, various techniques have been studied to provide more accurate predictions and recommendations to the active user. The common methods of recommendation are collaborative filtering (CF), content-based filtering, and hybrid techniques [2, 6–8]. CF technique selects an affinity group based on the preferences of active user. Thereafter, recommendations are generated by scrutinizing the preferences of users similar to the active user. The CF-based recommendation techniques are widely adopted and efficient techniques. Still, cold start problem, sparsity, scalability, and synonymy are the major issues associated with CF which limit the recommendation efficiency, whereas content-based recommendation technique suggests the relevant items to the active user based on content analysis. Limited content analysis and new item problems are the major concerns of these recommendation tech-

S. Tyagi · P. Yadav (✉) · M. Arora · P. Vashisth
Shyama Prasad Mukherji College, University of Delhi, Delhi 110026, India
e-mail: pratibha89yadav@gmail.com

S. Tyagi
e-mail: shwetakaushik2006@gmail.com

M. Arora
e-mail: moniarora0905@gmail.com

P. Vashisth
e-mail: poojamohitvashisth@gmail.com

© Springer Nature Singapore Pte Ltd. 2019
R. K. Shukla et al. (eds.), *Data, Engineering and Applications*,
https://doi.org/10.1007/978-981-13-6347-4_3

23

niques. Hybrid recommendation techniques are the fusion of the content-based and collaborative techniques to overcome the problems of recommendation techniques and to generate more precise recommendations. The proposed CF-based scheme adopts a regression model based on extreme learning machine (ELM) [9] to tackle the issues of sparsity and scalability.

The remainder of the paper is arranged as discussed below. Section 2 concisely encapsulates the CF-based RSs and working architecture of ELM. A novel ELM-based collaborative filtering model is deliberated in Sect. 3. Experimental results are conducted on MovieLens dataset which are explained in Sect. 4. Lastly, in Sect. 5, conclusions and future work are emphasized.

## 2   Background

With the advancement in the area of recommender systems, CF-based recommendation technique emerged as one of the most dominant recommendation techniques [10–13]. For making recommendations to the active user, this technique explores the interest of like-minded users. Due to sparsity in dataset, to select the group of similar users becomes inefficient. Further, the filtering process becomes time-consuming with an increasing number of users and items. Consequently, these challenges offer the task for the researchers for making more desirable and accurate recommendations.

In the literature, various regression algorithms have been suggested to handle the issues associated with CF-based recommendation engine. These algorithms are employed widely in the domain of CF and generate the quality recommendations for the user [10, 14–16]. In this work, ELM is adopted on the transformed dataset in order to tackle the problems of sparsity and scalability and to enhance the recommendation quality as well.

Originated from the single-hidden layer feed-forward neural network (SLFN), ELM is proposed by Huang et al. [17]. The ELM algorithm for SLFNs learns from $N$ training samples, $\{(x_i, t_i)|i = 1, 2, \ldots, N\}$, where $x_i = (x_{i,1}, x_{i,2}, \ldots, x_{i,p}) \in R^p$ and $t_i = (t_{i,1}, t_{i,2}, \ldots, t_{i,q}) \in R^q$. The working architecture of ELM, consuming $\widetilde{N}$ hidden neurons such that $\widetilde{N} \ll N$, is demonstrated in Fig. 1.

The standard SLFNs are modeled by [9] as given by the following mathematical formula:

$$\sum_{i=1}^{\widetilde{N}} \beta_i f_a(w_i \cdot x_j + b_i) = o_j, \quad j = 1, 2, \ldots N, \tag{1}$$

where $f_a$ symbolizes the activation function, $w_i$ represents the input weight vector, $\beta_i$ denotes the output weight vector, and $b_i$ symbolizes the threshold of the $i$-th hidden node [9, 17]. $w_i \cdot x_j$ indicates the inner product of $w_i$ and $x_j$. The above $N$ equations (for $j = 1, 2, \ldots, N$) can be represented as [18]:

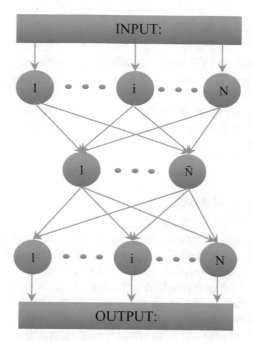

**Fig. 1** Working of extreme learning machine

$$H\beta = T \qquad (2)$$

where, $H$ symbolizes the hidden layer output matrix [19, 20]. The solution of Eq. (2) can be calculated by the following equation suggested by [17]:

$$\widehat{\beta} = H^{\dagger}T \qquad (3)$$

where $H^{\dagger}$ denotes the Moore–Penrose generalized inverse [20] of matrix $H$. After learning the value of $\widehat{\beta}$, the regression estimation value matrix $Y$ can be computed by:

$$Y = H\widehat{\beta} = HH^{\dagger}T. \qquad (4)$$

In this work, ELM is employed in the area of CF to deal with sparsity, improve the scalability, and enhance the accuracy of recommendations.

## 3   ELM-Based CF Model

The proposed ELM-based CF model is designed to handle the problems of sparsity and scalability as well as to generate desired recommendations. The design of the proposed approach is illustrated in Fig. 2.

For this purpose, the sparse rating dataset is transformed into a reduced-size dense dataset. On the transformed dataset, ELM is applied to make the model to learn the prediction of ratings.

### 3.1   *Reduction of Dataset*

The MovieLens dataset is studied to develop the proposed model. This dataset contains ratings of $n$ items (movies) given by $m$ users. Movies in the dataset are classified according to 18 genres. The rating matrix is denoted by $R_{m \times n} = [r_{i,j}]$, where $r_{i,j}$ denotes the rating of $j$th item provided by $i$th user. To handle the problem of sparsity and scalability together, the rating dataset is reduced to a dense dataset by capturing the users' interests in terms of genre interestingness measure (GIM) [21] as shown in Fig. 3. The GIM of $i$th user in $k$th genre is represented by $g_{i,k}$ and computed by employing the formula suggested by Mohammad and Bharadwaj [21]. Thus, computed dense dataset of GIM is denoted by $G_{m \times 18} = [g_{i,k}]$.

The main advantage of the reduction step is to handle the sparsity of dataset. Sparsity in rating matrix prevents the CF-based RS to select the appropriate set of like-minded users. In spite of having similar taste in genres, two users would not be identified similar if they have not explored common items. This issue can be resolved by capturing these users' interests in terms of GIM. Another advantage of the reduction step is to scale the number of items to the number of genres.

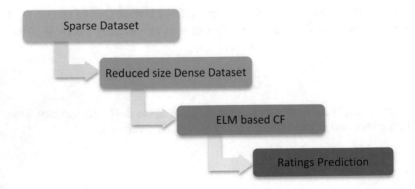

**Fig. 2**  Architecture of ELM-based CF model

| Rating Matrix $R_{m \times n}$ | | | |
|---|---|---|---|
| $r_{1,1}$ | $r_{1,2}$ | ... | $r_{1,n}$ |
| $r_{2,1}$ | $r_{2,2}$ | ... | $r_{2,n}$ |
| ... | ... | ... | ... |
| $r_{m,1}$ | $r_{m,2}$ | ... | $r_{m,n}$ |

Computation of GIM

| GIM Matrix $G_{m \times 18}$ | | | |
|---|---|---|---|
| $g_{1,1}$ | $g_{2,1}$ | ... | $g_{1,18}$ |
| $g_{2,1}$ | $g_{2,2}$ | ... | $g_{2,18}$ |
| ... | ... | ... | ... |
| $g_{m,1}$ | $g_{m,2}$ | ... | $g_{m,18}$ |

**Fig. 3** Transformation of rating dataset into GIM dataset

## 3.2 ELM for Rating Prediction

To understand the working of the proposed GIM and ELM-based CF algorithm, namely GECF, let $P_{p \times q} = [p_{i,j}]$, where $p_{i,j}$ symbolizes the rating of $j$th item predicted for the $i$th active user by the recommendation model, $\widetilde{N}$ represents the number of hidden nodes, and $f_a(x) = \frac{1}{1+e^{-x}}$ is the activation function [9]. The algorithm is designed as follows:

---

**Algorithm: GECF**

---

**Input:** $R_{m \times n}$, $GT_{m \times 18}(GIM\ for\ training\ users)$, $GA_{p \times 18}(GIM\ for\ active\ users)$

**Output:** $P_{p \times q}$

**Step 1:** Allocate bias $B_{1 \times \widetilde{N}}$ and input weight $W_{1 \times \widetilde{N}}$ randomly

**Step 2:** Compute the matrix of hidden layer $H$ using $GT_{m \times 18}$, $W$ and $B$ as follows:

$$H\left(W_{1 \times \widetilde{N}}, B_{1 \times \widetilde{N}}, GT_{1 \times \widetilde{N}}\right) = \left[ f_a(W_1 \cdot g_1 + B_1) \ldots f_a\left(W_{\widetilde{N}} \cdot g_1 + B_{\widetilde{N}}\right) \vdots \ldots \right.$$

$$\left. \vdots f_a(W_1 \cdot g_M + B_1) \ldots f_a\left(W_{\widetilde{N}} \cdot g_M + B_{\widetilde{N}}\right) \right]_{M \times \widetilde{N}}$$

**Step 3:** Compute the output vector $\widehat{\beta} = H^\dagger R$ where $H^\dagger$ is computed by employing the formula $H^\dagger = \left(H^T H\right)^{-1} H^T$ [20]

**Step 4:** Compute $H_{active}$ for the active users with $GA_{p \times 18}$, $W$ and $B$

**Step 5:** Compute the output using $P = H_{active} * \widehat{\beta}$

---

The proposed novel GECF algorithm computes GIM matrix for training and active users separately. Besides using the sparse rating matrix, GECF scheme employs the dense and reduced-size GIM matrix to learn the parameter $\beta$ and thereafter the ratings of items are predicted for each active user to generate recommendations.

## 4  Experimental Evaluation

This segment demonstrates the recommendation accuracy of the proposed approach against the state-of-the-art techniques. To authenticate the efficacy of the novel technique, various experiments are conducted on the real dataset. This section first describes the dataset adopted. Later, it explains the evaluation metrics used for the evaluation, and then, empirical results have been shown in the following sections.

### 4.1  Dataset

In this work, MovieLens dataset has been chosen for measuring the recommendation quality of the novel technique. MovieLens is a widely adopted movie dataset which comprises 1 million ratings for the 4000 movies given by 6000 registered users [22]. The ratings are on the scale of 1 (very bad) to 5 (very good). The dataset has a high level of sparsity because the movies have not been rated by the significant amount of users.

### 4.2  Evaluation Metrics

In recommender systems literature, researchers have proposed plentiful metrics to compute the efficacy of the collaborative filtering techniques [23]. For computing the classification and prediction accuracy of the novel technique, we have adopted various metrics.

#### 4.2.1  Classification Accuracy Measures

To examine the classification accuracy, the measures adopted are: precision, recall also known as true-positive rate (TPR), F-measure, true-negative rate (TNR), false-positive rate (FPR), and false-negative rate (FNR). For computing these measures, each item is categorized as significant or insignificant on the basis of the classification presented in Table 1.

**Table 1**  Classification of item

|               | Suggested to user      | Not suggested to user   |
|---------------|------------------------|-------------------------|
| Significant   | True positives (TPs)   | False negatives (FNs)   |
| Insignificant | False positives (FPs)  | True negatives (TNs)    |

Precision aims at finding the fraction of good items out of total suggested items. Recall focuses on the proportion of the noteworthy items recommended out of the total significant items. F-measure combines the usefulness of both the previously described metrics into one. The formula for the computation of F-measure is given below:

$$\text{F-measure} = \frac{2 * \text{Precision} * \text{Recall}}{\text{Precision} + \text{Recall}} \tag{5}$$

where precision and recall are evaluated using equations as follows:

$$\text{Precision} = \frac{\text{TP}}{\text{TP} + \text{FP}} \tag{6}$$

$$\text{Recall(TPR)} = \frac{\text{TP}}{\text{TP} + \text{FN}} \tag{7}$$

The equations for FPR, TNR and FNR are given below:

$$\text{FPR} = \frac{\text{FP}}{\text{FP} + \text{TN}} \tag{8}$$

$$\text{TNR} = \frac{\text{TN}}{\text{FP} + \text{TN}} \tag{9}$$

$$\text{FNR} = \frac{\text{FN}}{\text{TP} + \text{FN}} \tag{10}$$

The results computed using classification accuracy measures are discussed later.

### 4.2.2 Prediction Accuracy Measures

The accuracy of prediction is analyzed by computing mean absolute error (MAE) and root-mean-square error (RMSE). The absolute difference of the anticipated and original ratings is captured through the computation of MAE. RMSE is calculated by the square root of the mean of squared difference between the anticipated and original ratings. The formula for computing MAE and RMSE are given below:

$$\text{MAE} = \frac{\left| r_{u,i} - p_{u,i} \right|}{\text{total recommendations}} \tag{11}$$

$$\text{RMSE} = \sqrt{\frac{(r_{u,i} - p_{u,i})^2}{\text{total recommendations}}} \tag{12}$$

where $r_{u,i}$ denotes the actual rating of item $i$ by user $u$ and $p_{u,i}$ represents the predicted rating of item $i$ for user $u$.

## 4.3   Empirical Results

The proposed technique, GIM and ELM-based CF named as GECF, is examined against the state-of-the-art techniques, namely RCF [9] and PCF [3]. A chain of experiments has been carried out to show the evident soundness of the suggested techniques over the state-of-the-art approaches. For the sake of experimentation, 1000 training users and 100 test (active) users have been randomly extracted from the original dataset. In addition, for each test (active) user, we have divided the items into training and test items. For the selection of test items, initially we have randomly extracted five items to comprise the test set denoted by Test_5 and rest of the items form the training set of items. Likewise, the test set has been expanded on a scale of 5 such as Test_10, Test_15, and so on to test the prediction and classification accuracy of the novel technique in comparison with state-of-the-art techniques.

First, the proposed model has been analyzed against RCF on the basis of classification metrics. The results are shown in Table 2, which clearly reveal that the proposed method outperforms the RCF technique for all samples of test items.

In addition, to examine the remarkable classification capacity of the novel technique, the approaches have been assessed on the basis of TPR, FPR, TNR, and FNR. Consequently, the average values of TNR, TPR, FPR, and FNR are computed after each experiment and are depicted in Fig. 4.

Further, for comparison against PCF, we have adopted most notable MAE and RMSE metrics aforementioned in the previous section. The outcomes are illustrated in Table 3. The results obtained visibly indicate that the proposed technique outperforms the state-of-the-art techniques.

**Table 2** Performance comparison of GECF and RCF on the basis of recall, precision, and F-measure

| Test_items | Technique | Recall | Precision | F-measure |
|------------|-----------|--------|-----------|-----------|
| Test_5     | GECF      | 0.9234 | 0.8155    | 0.8662    |
|            | RCF       | 0.8476 | 0.7418    | 0.7912    |
| Test_10    | GECF      | 0.9581 | 0.8112    | 0.8786    |
|            | RCF       | 0.8411 | 0.7258    | 0.7792    |
| Test_15    | GECF      | 0.9448 | 0.8065    | 0.8703    |
|            | RCF       | 0.8570 | 0.7077    | 0.7752    |
| Test_20    | GECF      | 0.9672 | 0.8167    | 0.8857    |
|            | RCF       | 0.8565 | 0.8037    | 0.8293    |
| Test_25    | GECF      | 0.9727 | 0.8109    | 0.8845    |
|            | RCF       | 0.8512 | 0.7696    | 0.8084    |

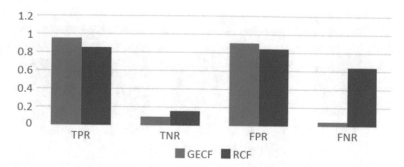

**Fig. 4** Average value of TPR, TNR, FPR, and FNR for GECF and RCF

**Table 3** Performance comparison of GECF and PCF on the basis of MAE and RMSE

| Test_items | Technique | MAE | RMSE |
|---|---|---|---|
| Test_5 | GECF | 0.8066 | 0.9138 |
| | PCF | 0.8529 | 0.9682 |
| Test_10 | GECF | 0.7594 | 0.9009 |
| | PCF | 0.8087 | 0.9165 |
| Test_15 | GECF | 0.7773 | 0.9082 |
| | PCF | 0.8218 | 0.9355 |
| Test_20 | GECF | 0.7290 | 0.8813 |
| | PCF | 0.7742 | 0.9071 |
| Test_25 | GECF | 0.7057 | 0.8746 |
| | PCF | 0.7419 | 0.8936 |

## 4.4 Selection of Parameter β (Beta)

In order to analyze the prediction efficiency of the novel technique, the beta parameter is trained. For the selection of $\beta$ parameter, 50 iterations of the algorithm are performed. In each iteration, two parameters, named max_count and min_count, are maintained. The max_count and min_count represent the maximum number and minimum number of nonzeros values in the beta parameter obtained after each iteration, respectively. At the end of the iterations, resultant two values of $\beta$ parameter are achieved corresponding to the least number of nonzeros values and maximum number of nonzeros values. For both values of $\beta$ parameter obtained on the basis max_count and min_count, F-measure of the proposed technique is computed to compare and analyze the recommendation ability. The comparison of F-measure of the two variants of the proposed technique is shown in Fig. 5.

The results depicted in Fig. 5 clearly expose that the variant of the proposed technique with the parameter $\beta$ corresponding to max_count performs better as compared with the other variants of the proposed technique with the parameter $\beta$ corresponding to min_count.

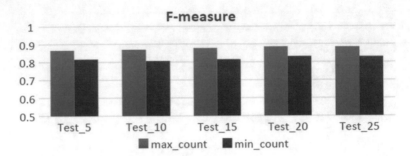

**Fig. 5** Comparison of F-measure for GECF on the basis of *max_count* and *min_count*

## 5 Conclusion

In this paper, a novel extreme learning machine-based model named GECF is proposed which utilizes the genre interestingness measure of the users to enhance the quality of the traditional ELM by alleviating the problem of the sparsity and scalability. Experiments conducted on the real dataset demonstrate that the proposed GECF scheme outperforms the state-of-the-art CF techniques in terms of prediction and classification accuracy as well.

Further, to significantly enhance the prediction quality of the novel technique, the $\beta$ parameter is trained through several iterations. In the future, we tend to enhance the performance of the algorithm by embracing genetic algorithm for the training of $\beta$, weight, and bias [24]. Furthermore, to tackle the issue of big data in the domain of recommender system, we aim at designing the parallel computing environment for the recommendation algorithm using Hadoop MapReduce framework [25].

## References

1. Adomavicius, G., Tuzhilin, A.: Toward the next generation of recommender systems: a survey of the state-of-the-art and possible extensions. IEEE Trans. Knowl. Data Eng. **17**(6), 734–749 (2005)
2. Bobadilla, J., Ortega, F., Hernando, A., Gutierrez, A.: Recommender systems survey. Knowl. Based Syst. 46, 109–132 (2013)
3. Resnick, P., Iakovou, N., Sushak, M., Bergstrom, P., Riedl, J.: GroupLens: an open architecture for collaborative filtering of netnews. In: Proceedings of 1994 Computer Supported Cooperative Work Conference, pp. 175–186 (1994)
4. Shardanand, U., Maes, P.: Social information filtering: algorithms for automating 'Word of Mouth'. In: Proceedings Conference on Human Factors in Computing Systems, pp. 210–217 (1995)
5. Sarwar, B., Karypis, G., Konstan, J.A., Riedl, J: Analysis of recommendation algorithms for e-commerce. In: Proceedings of the ACM E-Commerce, pp. 158–167 (2000)
6. Basu, C., Hirsh, H., Cohen, W.: Recommendation as classification: using social and content-based information in recommendation. In: Proceedings of the 15th National Conference on, Artificial Intelligence, pp. 714–720 (1998)

7. Tyagi, S., Bharadwaj, K.K.: A collaborative filtering framework based on fuzzy case-based reasoning. In: Proceedings of the International Conference on Soft Computing for Problem Solving, Springer AISC Series, vol. 130, pp. 279–288. Print ISBN: 978-81-322-0486-2, Online ISBN: 978-81-322-0487-9 (2012)
8. Claypool, M., Gokhale, A., Miranda, T., Murnikov, P., Netes, D., Sartin, M.: Combining content-based and collaborative filters in an online newspaper. In: Proceedings of the ACM SIGIR'99 Workshop on Recommender Systems (1999)
9. Deng, W., Zheng, Q., Chen, L.: Real-time collaborative filtering using extreme learning machine. In: IEEE/WIC/ACM International Conference on Web Intelligence and Intelligent Agent Technology—Workshops (2009)
10. Su, X., Khoshgoftaar, T.M.: A survey of collaborative filtering techniques. Adv. Artif. Intell. 3, 1–20 (2009)
11. Tyagi, S., Bharadwaj, K.K.: Enhancing Collaborative Filtering Recommendations by Utilizing Multi-objective Particle Swarm Optimization Embedded Association Rule Mining. Swarm and Evolutionary Computation. Elsevier, Netherlands, vol. 13, pp. 1–12. ISSN: 2210-6502 (2013)
12. Tyagi, S., Bharadwaj, K.K.: A particle swarm optimization approach to fuzzy case-based reasoning in the framework of collaborative filtering. Int. J. Rough Sets Data Anal. (IJRSDA) 1(1), 48–64. ISSN: 2334-4598, EISSN: 2334-4601 (2014)
13. Pennock, D.M., Horvitz, E., Lawrence, S., Giles, C.L.: Collaborative filtering by personality diagnosis: a hybrid memory and model-based approach. In: Proceedings of 16th Conference on Uncertainty in Artificial Intelligence, pp. 473–480 (2000)
14. Yu, C., Xu, J., Du, X.: Recommendation algorithm combining the user-based classified regression and the item-based filtering. In: Proceedings of the 8th International Conference on Electronic Commerce, ACM, pp. 574–578 (2006)
15. Zhu, H., Luo, Y., Weng, C., Li, M.: A collaborative filtering recommendation model using polynomial regression approach. In: The 4th China Grid Annual Conference, IEEE, pp. 134–138 (2009)
16. Chang, T., Hsiao, W., Chang, W.: An ordinal regression model with SVD Hebbian learning for collaborative recommendation. J. Inf. Sci. Eng. (Forthcoming)
17. Huang, G., Zhu, Q., Siew, C.: Extreme learning machine: a new learning scheme of feedforward neural networks. In: Proceedings of IEEE International Joint Conference on Neural Networks, vol. 2, pp. 985–990 (2004)
18. Huang, G.B., Zhu, Q.Y., Siew, C.K.: Extreme learning machine: theory and applications. Neurocomputing 70, 489–501 (2006)
19. Huang, G.B.: Learning capability and storage capacity of two-hidden-layer feedforward networks. IEEE Trans. Neural Netw. 14(2), 274–281 (2003)
20. Serre, D.: Matrices: Theory and Applications. Springer, New York (2002)
21. Al-Shamri, M.Y.H., Bharadwaj, K.K.: Fuzzy-genetic approach to recommender systems based on a novel hybrid user model. Expert Syst. Appl. 35(3), 1386–1399 (2008)
22. Miller, B., Albert, I., Lam, S., Konstan, J., Riedl, J.: MovieLens unplugged: experiences with an occasionally connected recommender system. In: Proceedings of ACM 2003 International Conference on Intelligent User Interfaces, ACM, pp. 263–266 (2003)
23. Herlocker, J.L., Konstan, J.A., Terveen, L.G., Riedl, J.T.: Evaluating collaborative filtering recommender systems. ACM Trans. Inf. Syst. 22(1), 5–53 (2004)
24. Bobadilla, J., Ortega, F., Hernando, A., Alcalá, J.: Improving collaborative filtering recommender systems results and performance using genetic algorithms. Knowl. Based Syst. 24(8), 1310–1316 (2011)
25. Dean, J., Ghemawat, S.: MapReduce: simplified data processing on large cluster. Commun. ACM 51(1), 107–113 (2008)

# Application of Community Detection Technique in Text Mining

Shashank Dubey, Abhishek Tiwari and Jitendra Agrawal

## 1 Introduction

The data mining is a technique that enables us to evaluate or analyze the data automatically. This ability of data mining offers to analyze the significant amount of data
in less amount of human effort consumption [1]. In this work, the data mining technique is utilized for text mining. The text mining is a domain where the data mining
algorithms and techniques are applied for analyzing the text data [2]. The text mining
techniques includes the various phases of data analysis such as pre-processing, feature computation, implementation of data mining algorithm on feature set recovered,
and finally the evaluation of performance on the basis of application [3].

In this context, various techniques of data mining techniques are developed that
are claimed to provide the accurate and efficient data analysis. But the text data is
not completely separable from each other; a partial similarity always exists among
various subjects of data or different domains of data. Therefore in order to understand
the similarity and the differences among two given text documents, they are needed
to be evaluated. In addition to that the text is an unstructured kind of data which is
not available in pre-labeled format. Thus, making accurate classification technique
for the text is also a complicated task [4].

In order to deal with the considered issues and text mining challenges, a new
approach of text mining is proposed in this work. That automatically analyzes the
data and discovers the possible similar groups in data. This automatic recovery of

S. Dubey (✉) · A. Tiwari
Information Technology, Mahakal Institute of Technology, Ujjain, India
e-mail: shashankdubey9@gmail.com

A. Tiwari
e-mail: abhi.tiwari23@gmail.com

J. Agrawal
Computer Science and Engineering, Rajiv Gandhi Proudyogiki Vishwavidyalaya, Bhopal, India
e-mail: jitendra@rgtu.net

© Springer Nature Singapore Pte Ltd. 2019
R. K. Shukla et al. (eds.), *Data, Engineering and Applications*,
https://doi.org/10.1007/978-981-13-6347-4_4

data groups is termed here as the community detection in the text mining. In addition to that to make understandable the similarity in interclusters [5], the visualization technique is used. In addition to that sometimes the two clusters can share some kind of common data, thus it is also considered to evaluate the common amount and instances of data among multiple groups. This section provides the basic overview of the proposed work. The next section describes the complete system modeling and the phases of the system.

## 2 Proposed Work

This section provides the detailed discussion and understanding about the proposed methodology. The methodology involves the solution development components, their functions, and the evaluation processes.

### 2.1 System Overview

The proposed work is motivated to perform text mining using the new technique which promises to not only provide accurate cluster analysis but also provide the relativity among the available clusters on to another. Basically, the cluster technique is an unsupervised manner of data mining. The unsupervised learning needs to provide the predefined patterns as input for learn about the patterns. These algorithms are developed in such manner by which the algorithm self-evaluates the data and finds the similarity or differences among them to prepare the similar kind of data objects or instances.

The text mining technique is now in these not only helpful for performing the classical task such as digital document classification and categorization. That is also much effectively used for various other applications such as finding the trending topics in social media, obtaining the positive and negative reviews of company products and services, emotion mining, and similar others [6]. Therefore, the text mining using the unsupervised learning is a subject of interest in this work. The proposed text mining technique is motivated from a community detection approach. Basically, the community detection approach is a problem-solving method which uses the graph theory for finding the similar group of objects in a significant amount of data.

In this work, the key method utilizes the text data as input and analyzes the text using distance finding method. Finally, the relativity among the data is demonstrated using the graph method. The main issue during this process is to regularize the size of text data and the recovery of text features. These features are the properties by which the similarity or difference among two data instances is measured. This section provides the overview of the proposed model, and the next section provides the detailed description of model formulation.

## 2.2 Methodology

The process used in the proposed methodology for community detection in text data is demonstrated in Fig. 1. In this diagram, the block contains the processes and the edge of blocks represents the flow of processed data from one process to another.

**Input text data**: That is an initial input to the system. The experimenting user can use any kind of text data using this input provision. But the limitation is only that the input data is needed to be combined in a single text document or a single directory. The system read this input document or set of document to utilize in further process. In order to collect the various domains or subjective data, here the social media text is used as input for experimentation purpose—more specifically the Twitter conversion (post) is used in a single document.

**Data pre-processing**: The pre-processing of data is one of the essential phases of data mining. The pre-processing technique is used to improve the quality of data. Therefore, the noisy attributes and instances are identified and removed from the input datasets. In the context of text document, the noise is recognized as the unused data, words, symbols which appeared in document or text blocks and not having much importance in document orientation discovery. Therefore, the proposed system includes two phases of data pre-processing.

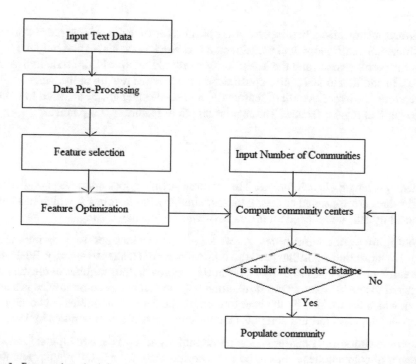

**Fig. 1** Proposed methodology

1. Removing the symbols from the social media text blocks, these symbols include the special characters and unwanted symbols. In order to define this task, a user-defined list of unwanted symbols is prepared and provided as input to the system. System picks one word at a time and removes it from the entire input text data.
2. In this phase, the similar function of the data replacement is used as defined in the first pre-processing stage. In this phase in place of special character list, the list of words that is known as noisy words are included in list. Additionally, using the similar function the words from the entire input text dataset are removed.

**Feature selection**: The input text dataset is reduced after pre-processing of the data. Therefore, in further process the dataset is analyzed in order to find the important or highly weighted terms. Thus, the probability of each word is computed which is used in dataset. In this context, the word frequency is one of the most essential feature computation techniques that provide the probability of individual word in the given domain. The word frequency $f$ can be computed by the ratio of total time a word appeared in document $W_c$ or total count of words in document and the total words available in document $W_t$. Thus, word frequency is given by [7]:

$$f = \frac{W_c}{W_t}$$

Thus in this phase, the words are evaluated in order to find the importance of the word.

**Feature optimization**: In this phase, the optimization of features is performed during optimization; each individual social media text block is evaluated and only the high priority words remain, and the less priority words are removed from each individual posts. In addition to that it also considers all the features remain in the same length. Therefore, the largest length of features $F_l$ and smallest length $F_s$ are used to define the length of regular feature. The regular length of feature is computed as:

$$R = \frac{F_l + F_s}{2}$$

where $R$ is length of each feature. The feature list which does not have enough data, null values are appended at their end, and similarly the feature list which has large amount of data are reduced on the basis of word frequency sorting.

**Input number of communities**: As we know that any unsupervised learning technique or clustering algorithm needs an input number of clusters to identify. Basically, the entire process leads to design a clustering approach, thus number of clusters are required to process data. In the similar manner, the number of communities is a user input which is provided by the user to identify the data communities. According to the user's input, algorithm computes the different possible communities in data.

**Compute community centers**: That phase initially accepts the two inputs: First, the number of communities needs to be detected and second the list of regular features. Initially, the similar number of instances of data is selected from the feature list as number of communities is needed to detect randomly. These selected instances of features are termed here as the community centers. Now each individual community center is compared with the all the list of data instances. In order to compare the community center and instance of data, two distance functions are used.

1. **Word-to-word comparison**: In order to compare and find the difference among two instance features, the Levenshtein distance [8] is used. That results the amount of characters similar to each other.
2. **Combined difference**: After computing the difference between two data instances, the obtained dissimilarity is converted into the percentage value. Because Levenshtein distance values can be higher than the 100 points, thus to regulate the values the percentage conversion is performed.

Based on the obtained distance among the points (selected community center and the data instance), the community is prepared which is having maximum similarity. Therefore, the similarity is given by:

$$\text{Sim} = 100 - \text{difference}$$

After computing the first initial phase of community, the optimization of the community centers is required, by which the dense and more effective communities are identified. In order to obtain this, combined similarity score or the internal community distance is computed by the following formula:

$$\text{ID} = \frac{1}{N} \sum_{i=1}^{N} \text{Sim}_i$$

If the ID values of the community centers are in a range, then the optimization process is stopped, otherwise the previous community centers are again computed. During computation of the new community centers, the intersection of nearest points is used to replace the previous community center, and again the process of similarity measurement is performed. This process is carried out till all the IDs remain in a range or the numbers of default evaluation cycles are reached.

**Populate community**: That is the final phase of modeling where the computed communities are visualized using the graphical manner. That visualization helps to provide understanding about the instances of data which belongs to the same community. In addition to that it also provides the information which data is partially similar between more than on communities. In addition to that the performance of the system is also computed in this phase for recognizing the efficiency and accurate cluster formation ability.

This section provides the detailed understanding about the proposed community detection model. In next section, the algorithm of the system is provided.

**Table 1** Proposed algorithm

| |
|---|
| Input: Text dataset D, Number of communities C, Optimization rang R |
| Output: number of communities groups G |

Process:

1. $R = readDataset(D)$

2. $P = PreProcessData(R)$

3. $T_N = TokenizeData(P)$

4. $for(i = 1; i \leq N; i + +)$

    a. $[f, T] = \frac{W_c}{W_t}$

5. $end\ for$

6. $Om = feature.Optimize(f, T, P)+$

7. $I_c = $ Select C instance from $Om+$

8. $for(j = 1; j \leq C; j + +)$

    a. $X = I_j$

    b. $for(k = 1; k \leq m; k + +)$

        i. $D = $ LevenshteinDistance$(X, O_k)$

        ii. $if(D < 0.25)$

            1. $G_j(I_j, O_k)$

        iii. $end\ if$

    c. End for

9. End for

10. $for(l = 1; l \leq C; l + +)$

    a. $ID_l = \frac{1}{N}\sum_{i=1}^{N} Sim_i$

    b. $if(ID_l \leftarrow R)$

        i. Return $G_j(I_j, O_k)$

    c. Else

        i. Go to step7

    d. End if

11. End for

## 2.3 Proposed Algorithm

This section provides the summarized steps of the proposed methodology in terms of algorithm steps. The step process of the working is defined using Table 1.

# 3 Results Analysis

This section provides the evaluation of the proposed text mining technique using community detection approach. Therefore, different performance parameters are computed and their captured results are reported in this section.

## 3.1 Precision

Precision of the any algorithm demonstrates the part of data actually relevant to the current instance of community. That can be evaluated by the following formula:

$$precision = \frac{relevant\ pattern \cap total\ pattern}{total\ pattern}$$

The precision of the proposed technique of text mining is defined in Fig. 2. In this diagram, the $X$-axis indicates the amount of data instances used for experimentation and the $Y$-axis shows the obtained precision values between 0 and 1. According to the demonstrated results, the performance of the system improves with the amount of data instances for community identification. Therefore, the proposed model is acceptable for the utilizing with the different text mining application.

**Fig. 2** Precision

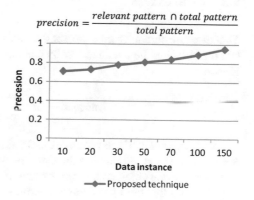

**Fig. 3** Recall

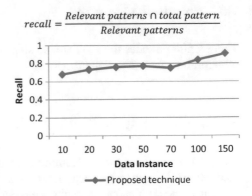

## 3.2 Recall

Recall is the amount of data that are recognized during the clustering process is relevant to be the same cluster. That can be estimated using the following formula:

$$recall = \frac{Relevant\ patterns \cap total\ pattern}{Relevant\ patterns}$$

The computed recall of the community detection-based text clustering or community detection system is described in Fig. 3. The $X$-axis of the diagram explains the amount of data instances consumed during the experimentation. Additionally, the $Y$-axis demonstrates the obtained fraction of recall values between 0 and 1. According to this parameter, the performance is acceptable due to continuous improvement of recall values for increasing amount of data instances.

## 3.3 F-measures

That measure combines precision and recall in terms of harmonic mean of precision and recall rate of the obtained results that can also be termed F-measure or balanced F-score:

$$F\text{-measure} = 2 \cdot \frac{precision * recall}{precision + recall}$$

The F-measures of the proposed text mining technique are given in Fig. 4. The $X$-axis of this figure contains the amount of experimental data supplied, and the $Y$-axis shows the obtained harmonic mean of precision and recall values. The result demonstrates that both the parameters are enhanced with the increasing amount of instances. Therefore, the proposed data model is acceptable for real-world text mining applications.

**Fig. 4** F-measures

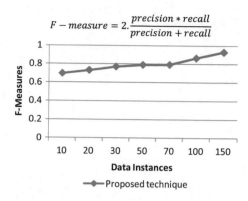

$$F - measure = 2.\frac{precision * recall}{precision + recall}$$

**Fig. 5** Time requirements

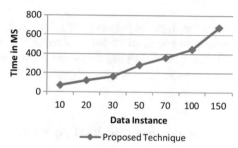

## 3.4 Time Requirements

The algorithms need a time to process the input data using the applied algorithms. The amount of time required for this purpose is termed here as time requirements or time complexity of the algorithm. The time requirement of the proposed technique is defined using Fig. 5. In this figure, X-axis indicates that the amount of dataset instances is produced for experiments and the Y-axis simulates the obtained time requirement. The time measurement is performed here in milliseconds. According to the obtained performance, the proposed model time consumption increases with the amount of data due to the number of optimization phases, in order to find optimal results. Thus, the model is acceptable for working with accurate data analysis.

## 3.5 Memory Usage

The processes need an amount of memory to be executing their task successfully. This amount of main memory is termed as memory usages or the space complexity of the algorithms.

The required amount of main memory is demonstrated using Fig. 6. The amount of main memory is described in Y-axis of the diagram, and the X-axis contains the

**Fig. 6** Memory usage

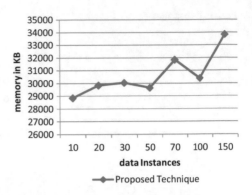

—◆—Proposed Technique

amount of data size provided input for processing. According to the obtained memory utilization graph, which is measured in KB (kilobytes); it increases as the amount of to be processed. But the nature of memory usages is not obtained in regular manner that produces uneven ups and downs.

# 4 Conclusion and Future Work

This section provides the summary of the entire work performed for designing and developing the proposed text clustering technique. The summary of the work is demonstrated here as the conclusion, and the limitations and possible future extension are described as the future work for the system.

## 4.1 Conclusion

The proposed technique is community detection technique for mining the text data. This technique used the text data and conducted analysis in three main phases: first the pre-processing or preparation of data and their format, second the feature selection and optimization, and finally implementation of cluster formation technique. The model is a continuous process of data analysis and optimization which can take a significant amount of time for completing the process of accurate data analysis. Thus, two additional limits or stopping criteria are developed. First criteria are to reach the number of iterations reached and second are if the objective function is accomplished during the clustering processes. Both the processes help to perform the accurate and dense clusters which additionally also provides the information about the shared data instances.

The implementation of the proposed technique is provided in JAVA technology. After the implementation, the experimental results are computed and on the basis of these experiments, the outcomes of the system are reported in Table 2.

**Table 2** Performance summary

| S. no. | Parameters | |
|--------|------------|---|
| 1 | Precision | Optimizes with the amount of data but becomes consistent after a significant amount of data |
| 2 | Recall | Optimizes the accuracy of recognition with amount of data |
| 3 | F-measures | Increases with the amount of data means provides more promising results |
| 4 | Time requirement | Increase with the amount of data but also depends upon the number of evaluation cycles |
| 5 | Memory usage | Uneven in nature and increases with the size of input data |

According to the obtained performance, the proposed technique is found acceptable for accurate data analysis.

## 4.2 Future Work

The proposed work is a promising approach for analyzing data that can be used in various complicated places where the pre-labeled data is not available; therefore, the following extensions are possible with the proposed technique.

1. The technique can be extended for social media topic tracking and trending topic detection
2. Providing good efforts for stream data mining techniques
3. Suitable to utilize the digital document retrieval system implementation

## References

1. He, W., Zha, S., Li, L.: Social media competitive analysis and text mining: a case study in the pizza industry. Int. J. Inf. Manage. **33**, 464–472 (2013)
2. Patel, R., Sharma, G.: A survey on text mining techniques. Int. J. Eng. Comput. Sci. **3**(5), 5621–5625. ISSN:2319-7242 (2014)
3. Mostafa, M.M.: More than words: social networks' text mining for consumer brand sentiments. Expert Syst. Appl. **40**, 4241–4251 (2013)
4. Aggarwal, C.C., Zhao, Y., Yu, P.S.: On the use of side information for mining text data. IEEE Trans. Knowl. Data Eng. **26**(6) (2014)
5. Kou, G., Peng, Y., Wang, G.: Evaluation of clustering algorithms for financial risk analysis using MCDM methods. Inf. Sci. **275**, 1–12 (2014)
6. Pang, B., Lee, L.: Opinion mining and sentiment analysis. Found. Trends Inf. Retrieval **2**(1–2), 1–135 (2008)

7. Nassirtoussi, A.K., Aghabozorgi, S., Wah, T.Y., Ngo, D.C.L.: Text mining for market prediction: a systematic review. Expert Syst. Appl. **41**, 7653–7670 (2014)
8. Michael Gilleland, Merriam Park Software: Levenshtein distance. In: Three Flavors. https://people.cs.pitt.edu/~kirk/cs1501/Pruhs/Spring2006/assignments/editdistance/Levenshtein%20Distance.htm

# Sentiment Analysis on WhatsApp Group Chat Using R

Sunil Joshi

## 1 Introduction

In today's internet era, the fastest communication through mobile application is WhatsApp, in which every smart phone user is sharing their thought, sentiment, and opinion with each other [1]. The main feature of this application is the group chat, through which a message can be easily accessed to many people, and that is why nowadays any news is spread by WhatsApp very quickly whether the news is constructive or unconstructive. Therefore, it has become very important to analyze each group's chat and find out how positive the group is. WhatsApp provides a facility to convert any group chat into a text file via e-mail but it is quite difficult task to analyze it manually. Sentimental analysis is a process by which the statement can be analyzed from large text and this process is also known as opinion mining [2]. R programming lab is a programming language that has many functions and packages available for text preprocessing, mining, and sentiment study. We can easily execute text mining and sentiment investigation by R Studio [3]. In this research paper, I reviewed the work done in the field of sentimental analysis research and has also reviewed the search work done on the data of WhatsApp. I have retrieved WhatsApp group data of all members of an organization and applied sentiment study and text mining with the help of R Studio. I have also analyzed the results of the level of emotions and sentiment of this WhatsApp group.

The rest of the paper is organized as follows: Sect. 2 discusses brief journalism of sentiment investigation and WhatsApp chat analysis. Section 3 focuses on implementation of sentiment investigation in R. Section 4 discusses about the results obtained for sentiment analysis. Finally, Sect. 5 concludes the research paper.

S. Joshi (✉)
Department of Computer Applications, Samrat Ashok Technological Institute,
Vidisha, MP, India
e-mail: sunil.joshi05@gmail.com

© Springer Nature Singapore Pte Ltd. 2019
R. K. Shukla et al. (eds.), *Data, Engineering and Applications*,
https://doi.org/10.1007/978-981-13-6347-4_5

## 2  Literature Review

Sentiment analysis or it is also known as opinion mining is a very trendy research topic in the field of big data analytics and data mining [4–7]. Lot of research works are done to analyze sentiment and opinion from various real-time big data like social networking site, micro-blogging site, mobile messaging applications, product review, movie review data, and many more [8–17]. As per [4], sentiment examination is a very critical task and it can be performed by five main steps which is presented in Fig. 1.1. The main categories of sentiment classification approaches are first machine learning, second lexicon based, and third hybrid approach [4]. There are various tools available like Emoticons, SentiWordNet, etc. for sentiment analysis [4]. One author performed experiment on 1000 posts from Facebook data and applied sentiment mining system [11].

WhatsApp provides a good solution for the problem of communication between two individuals as well as group of persons [18]. The most frequently used communication medium is Smilytext as compared to text or multimedia. The author reported that the most vigorous day of the week is Monday where maximum numbers of messages are transferred in WhatsApp group chat and the peak time where maximum messages are sent is after noon. WhatsApp is very popular mobile application worldwide; almost 750 million users are sending messages through WhatsApp and almost 20 million users are adding every month [19]. As per the survey, the maximum percentage of WhatsApp users are from age group from 26 to 35 and the minimum percentages are from senior citizen. WhatsApp provides a facility to send different types of messages like text, images, video, audio, and others. Author reported that text messages are used more as compared to other types of messages [19]. Michael surveyed the usage of WhatsApp group chat communication and their impact on network traffic [1]. One author found out the optimistic and pessimistic impacts on

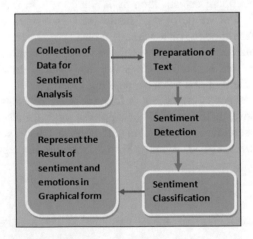

**Fig. 1.1**  Process of sentiment analysis

WhatsApp users using rapid minor tool [20]. Luca analyzes the twitter data about WhatsApp when WhatsApp service was stopped on the 3 May 2017 for 3 h.

The report stated that if one communication service is stopped, then the users will go to other communication service [21].

## 3 Implementation of Sentiment Analysis Using R Studio

The Implementation of sentiment analysis for WhatsApp group chat using R Studio having following steps:

1. **Retrieving Group Chat Data**: WhatsApp provides a facility to share any chat by e-mail. This option sends a text file to particular e-mail account. The steps to retrieved chat data using e-mail from WhatsApp are shown in Fig. 1.2.
2. **Installing R**: R is a programming lab and its developing environment is R Studio. R lab can be downloaded from its official website http://cran.r-project.org/ and

```
Open
WhatsApp   →   Open the
               Particular
               Group      →   Click on
                             the Menu
                             Button
                                 ↓
Select
Without Media  ←   Select
Option           Email chat  ←   Click on
                                 More
```

**Fig. 1.2** Extraction of WhatsApp chat data using e-mail

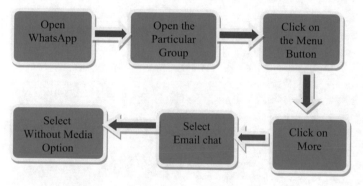

**Fig. 1.3** Installation of package for sentiment analysis

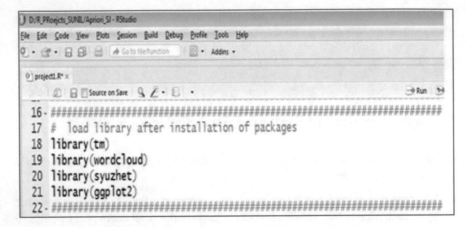

**Fig. 1.4** Load Library for sentiment analysis

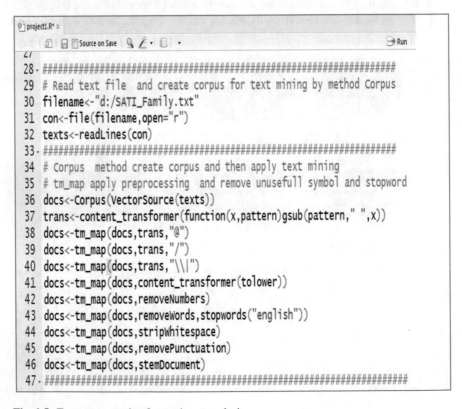

**Fig. 1.5** Text preprocessing for sentiment analysis

```
50 . ################################################################
51 #   after text mining wordcloud method create wordcloud with their frequency, for colour
52 # install package RColorBrewer
53 dtm<-TermDocumentMatrix(docs)
54 mat<-as.matrix(dtm)
55 v<-sort(rowSums(mat),decreasing = TRUE)
56 d<-data.frame(word=names(v),freq=v)
57 head(d,10)
58 set.seed(1056)
59 wordcloud(word=d$word,freq=d$freq,min.freq=1,max.words=200,randome.order=FALSE,
60         rot.per=0.35,colors=brewer.pal(8,"Dark2"))
61 . ################################################################
62
```

**Fig. 1.6**   Coding for creating word cloud

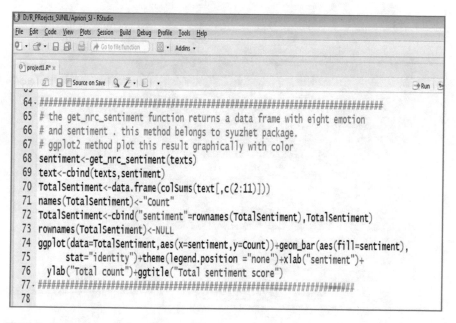

```
64 . ####################################################################
65 # the get_nrc_sentiment function returns a data frame with eight emotion
66 # and sentiment . this method belongs to syuzhet package.
67 # ggplot2 method plot this result graphically with color
68 sentiment<-get_nrc_sentiment(texts)
69 text<-cbind(texts,sentiment)
70 TotalSentiment<-data.frame(colSums(text[,c(2:11)]))
71 names(TotalSentiment)<-"Count"
72 TotalSentiment<-cbind("sentiment"=rownames(TotalSentiment),TotalSentiment)
73 rownames(TotalSentiment)<-NULL
74 ggplot(data=TotalSentiment,aes(x=sentiment,y=Count))+geom_bar(aes(fill=sentiment),
75      stat="identity")+theme(legend.position ="none")+xlab("sentiment")+
76   ylab("Total count")+ggtitle("Total sentiment score")
77 . ####################################################################
78
```

**Fig. 1.7**   Sentiment analysis coding

R Studio can be downloaded from https://download1.rstudio.org/RStudio-1.0.143.exe.

3. **R Package Installation**: R Studio environment provides a number of packages [22]. We need to install these packages first and then we are able to call any methods of these packages. Different types of packages are available for different types of functioning. For mining text and sentiment examination, we need to install these packages—tm, wordcloud, syuzhet, ggplot2, etc. The coding for installation of relevant package is shown in Fig. 1.3.

4. **Load Library**: After installation of relevant packages, we can load their library so that we are able to call their methods which are shown in Fig. 1.4.

5. **Read File**: Read text file which contains WhatsApp group chat database by method file and create text by readlines method.

6. **Preprocessing and Text Mining**: The package tm provides number of methods for text preprocessing and text mining which remove unwanted characters

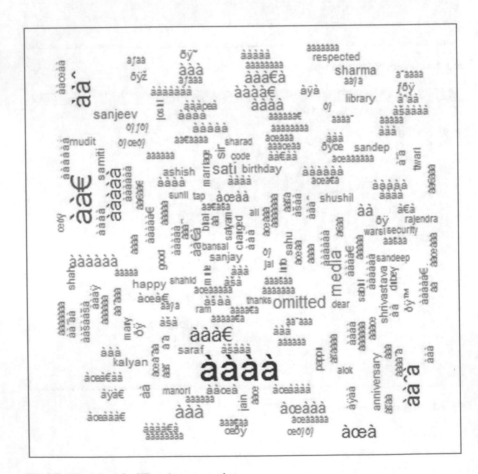

**Fig. 1.8** Wordcloud for WhatsApp group chat

and stop word and then apply text mining. These methods are Corpus, content_transformer, tm_map, etc. The coding of text mining in R is shown in Fig. 1.5.

7. **Construct WordCloud**: After converting data to frame the method, wordcloud creates cloud of word with their frequency which is shown in Fig. 1.6.
8. **Sentiment Analysis**: The package syuzhet provides method get_nrc_sentiment by which we can extract sentiment and eight emotions with their frequency and by ggplot2 method, we can present result in graphical format. The coding of sentiment investigation in R is shown in Fig. 1.7.

## 4 Result Analysis

The method wordcloud creates cloud of word with their frequency means the word and their count. The output of method wordcloud is presented in Fig. 1.8. If we observed the wordcloud then we found that which word is used most. In wordcloud method which words are most often used we can be seen by large font size. Now

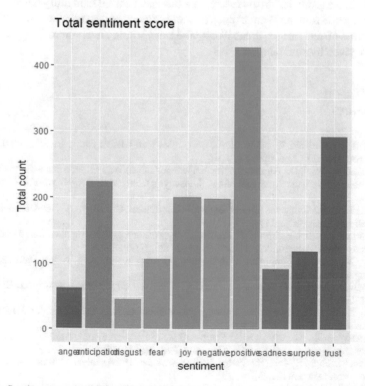

**Fig. 1.9** Sentiment result of WhatsApp group chat

for sentiment analysis, call method get_nrd_sentiment retrieves sentiment from text data. We can present eight emotions status in the documents graphically by method ggplot2. The result obtained from ggplot2 method is shown in Fig. 1.9. We observed that what level of sentiment used in this group chat. In this group, positive sentiment is so high and second largest sentiment is trust.

## 5    Conclusion

Today WhatsApp mobile application is the most popular mobile application of communication, but today its negative consequences are also in which any negative news is spread rapidly through WhatsApp and this problem is inspirational for us. I did my research work in the field of sentiment investigation and main intention was to investigate the chat of WhatsApp group and find out the intensity of sentiment and emotions in it. Many research works are done to analyze the use, addiction, and network effect of WhatsApp. In this sequence, it analyzed the group chat of all the members of an institution and determined the level of sentiment and emotion. Through the results I found "how positive the group is and how the level of eight emotions in the group is." In the future, this task can be used to identify negative users and negative ideas in the WhatsApp group and to analyze the statement of sentiments and emotions from various areas like social networking sites, micro-blogging sites, feedback given by students, etc.

## References

1. Seufert, M., Hoßfeld, T., Schwind, A., Burger, V., Tran-Gia, P.: Group-based communication in WhatsApp. IFIP Networking (2016)
2. Thakkar, H., Patel, D.: Approaches for sentiment analysis on Twitter: a state-of-art-study. In: International Network for Social Network Analysis Conference (INSNA), Xi'an China, July 2013
3. Torgo, L.: Data Mining with R Learning with Case Studies. CRC Press, Taylor & Francis Group an Informa Business (2011)
4. D'Andrea, A., Ferri, F., Grifoni, P., Guzzo, T.: Approaches, tools and applications for sentiment analysis implementation. Int. J. Comput. Appl. 125(3) (2015)
5. Fang, X., Zhan, J.: Sentiment analysis using product review data. J. Big Data (2015) (A SpringerOpen Journal)
6. Hussein, D.-M.E.D.M.: A survey on sentiment analysis challenges. J. King Saud Univ. Eng. Sci. (2016)
7. Joshi, K., Patel, D., Pandya, S.: A survey on sentiment analysis techniques. Int. J. Innov. Res. Technol. 3(7) (2016)
8. Jotheeswaran, J., Koteeswaran, S.: Sentiment analysis: a survey of current research and techniques. Int. J. Innov. Res. Comput. Commun. Eng. 3(5) (2015)
9. Kumar, S., Singh, P., Rani, S.: Study of difference sentimental analysis techniques: survey. Int. J. Adv. Res. Comput. Sci. Softw. Eng. 6(6) (2016)
10. Medhat, W., Hassan, A., Korashy, H.: Sentiment analysis algorithms and applications: a survey. Ain Shams Eng. J. 5 (2014)

11. Neri, F., Aliprandi, C., Capeci, F., Cuadros, M.: Sentiment analysis on social media. In: IEEE/ACM International Conference on Advances in Social Networks: Analysis and Mining (2012)
12. Pradhan, V.M., Vala J., Balani, P.: A survey on sentiment analysis algorithms for opinion mining. Int. J. Comput. Appl. **133**(9) (2016)
13. Safrin, R., Sharmila, K.R., Subangi, T.S., Vimal, E.A.: Sentiment analysis on online product review. Int. Res. J. Eng. Technol. **04**(04) (2017)
14. Shaikh, A., Rao, M.: Survey on sentiment analysis. In: International Conference on Emanations in Modern Technology and Engineering (ICEMTE-2017), vol. 5, issue 3 (2017)
15. Varghese, R., Jayasree, M.: A survey on sentiment analysis and opinion mining. Int. J. Res. Eng. Technol. **2**(11) (2013)
16. Vinodhini, G., Chandrasekaran, R.M.: Sentiment analysis and opinion mining: a survey. Int. J. Adv. Res. Comput. Sci. Softw. Eng. **2**(6) (2012)
17. Yadav, J.: A survey on sentiment classification of movie reviews. Int. J. Eng. Dev. Res. **3**(11) (2014)
18. Patil, S.: WhatsApp group data analysis with R. Int. J. Comput. Appl. **154**(4) (2016)
19. Deshmukh, S.: Analysis of WhatsApp users and its usage worldwide. Int. J. Sci. Res. Publ. **5**(8) (2015)
20. Radha, D., Jayaparvathy, R., Yamini, D.: Analysis on social media addiction using data mining technique. Int. J. Comput. Appl. **139**(7) (2016)
21. Whatsapp is Down, Now What? http://data-speaks.luca-d3.com/2017/05/whatsapp-is-down-now-what.html,Online. Access 10 June 2017
22. Zhao, Y.: R Reference Card for Data Mining. http://www.Rdatamining.com Online. Access 10 June 2017

# A Recent Survey on Information-Hiding Techniques

Jayant Shukla and Madhu Shandilya

## 1 Introduction

The Internet era started in late 1960s and 1970s to fulfil the requirement to enable communication in the battlefield to communicate critical information could prove beneficial in the war situations and to exchange research information between the researchers across different universities and different countries. Growth of Internet is very helpful in the field of communication. But the dark side of this growth (easy sharing of information) is unauthorized access of information and misuse of information has increased by time. So, the confidentiality and security of the susceptible information have been of greatest importance and subject of top priority.

Top reasons for this confidentiality and security are unreliability and less security of fundamental communication network over which the transfer of important information is carried out. Person who is having proper knowledge of network programming and having right application programs can eavesdrop and change communication information and also can intercept the data transfer which could be very unsafe.

So to overcome the security and authenticate access problem of sensitive information, there are many solution proposed like watermarking, steganography, etc. [1]. Basically watermarking and steganography both come under data-hiding techniques, where aim is to hide secret and sensitive information from unauthenticated peoples [2]. Digital watermarking concerns with hiding of information to the digital media itself. The embedding happens by changing the content of the digital media. Here, embedding of information is not done in the framework data. In digital watermarking

J. Shukla (✉)
MANIT, Bhopal, India
e-mail: jayantshukla81@gmail.com

M. Shandilya
Department of ECE, MANIT, Bhopal, India
e-mail: madhushandilya@manit.ac.in

© Springer Nature Singapore Pte Ltd. 2019
R. K. Shukla et al. (eds.), *Data, Engineering and Applications*,
https://doi.org/10.1007/978-981-13-6347-4_6

criteria is that the modifications of the media are unnoticeable. On the other hand, steganography is the most useful data-hiding technique in present days. Steganography is basically Greek word meanings covered or secret writing [3]. Hiding the presence of the message is the objective of steganography techniques, so that an eavesdropper or onlooker is unaware that the information is present or not.

Various applications create interest in the topic of information hiding [4].

(1) Security or intelligence agencies and military want inconspicuous transportation of information. Still as the data is coded, the discovery of presence of any doubtful signal in the current battle field possibly will soon result in the assault on the suspected. Due to this, military communication employs meteor scatter transmission or spread spectrum modulation techniques to make signals difficult for the opponent to detect.

(2) Law breakers give lot of importance to inconspicuous communications. Their favoured technologies comprise of those mobile phones that change their individuality frequently, prepaid mobile phones.

(3) Of let few governments have tried to restrict free dialogue on Internet as well as resident utilization of cryptography has urged the concerned peoples regarding autonomy to grow methods for unnamed communications on the Internet, together with Web proxies along with unnamed remailers.

(4) Digital cash and digital election schemes for utilizing secret communication methods.

(5) A person or company that are in advertising are using forgery techniques through email to send vast amount of unwanted messages at the same time avoiding responses from annoyed ones.

The whole paper is organized as follows. Initially, information hiding, like digital watermarking, steganography, and fingerprinting is discussed. Then, a broad range of methods that are used in a number of applications are described. Then, the attacks against all the methods are analysed; and lastly, attempt to devise principles and definitions.

## 1.1 Information Hiding

Information hiding concerns with hiding of information from unauthenticated person. This can be done by encryption of information or by some watermarking technique or by anonymity. There are many techniques by which information hiding can be done.

In Fig. 1 classification of some widely used information-hiding techniques is done. Here, anonymity can be viewed as a requirement for information hiding and mostly implemented in multi-agent system where goal is creating a condition of being anonymous [5]. The essential instinct behind anonymity is that actions should be separated from the agents who execute them, for a few set of observers. For example, the information which is required to be hidden is the identity of the agent/agents

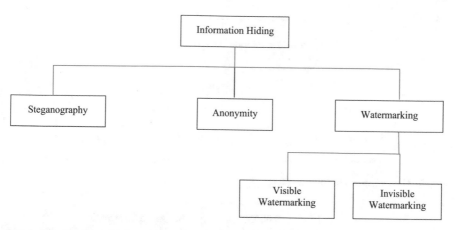

**Fig. 1** Classification of information-hiding technique

who carry out a particular action, whose information needs to be hidden from some observers (all unclassified agents). The third part of information-hiding requirements is how good information needs to be hidden. These three basic questions make a framework for anonymity [5].

The embedded data is the message which has to be sent secretly. The original message is concealed within another file and that file may be in the form of text, image or audio as per the requirement. A key called stego key is used to manage the hiding technique of the original data and to extract by the authenticate receiver. Steganography is a technique in which two users communicate with each other secretly and a successful attacker is the one who is unaware of the users but detects the communication.

Steganography is at times wrongly confused with cryptography, but there are some prominent and unique differences between them. In a few circumstances, steganography is favoured to the later one since in the cipher text it is a jumbled output of the plaintext and the invader can estimate that encryption is being done and therefore can utilize the methods of decryption to retrieve the hidden data. Cryptography methods also need higher level of computing power to carry out encryption which may imply severe obstruction for smaller devices which are short of sufficient computing resources to realize encryption.

Whereas, steganography is the method of protecting the important records in some cover media such as audio, video or images, on the Internet. Due to that the invader is not able to understand that the transmission of data is happening because data is hidden to the bare eye and impossible to separate from the original media. Here as shown in Fig. 2, cover media can be a text message, audio signal, or a digital image. Throughout the paper, digital image is considered as cover media. So, images generally employ 8-bit or 24-bit colour model. 8-bit colour model defines up to 256 colours making a palette for this image, every colour depicted by associate degree 8-bit value, example pure black can be represented by value 256 and pure white by

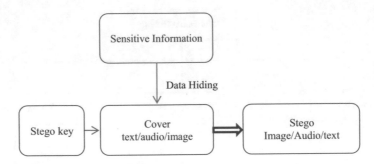

**Fig. 2** General model of steganography

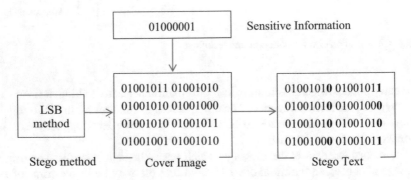

**Fig. 3** Illustration of simple steganography method

value 0 so 8 bit value of white colour is 000000 and for black colour it is 111111 and similarly other colours are defined between this gamut. 24-bit colour scheme uses twenty-four bits every pixel and gives higher set of colours. Here, the primary colour's (RGB) intensity is denoted by 3 bytes or 24 bits of every pixel. On the other end, secret or sensitive information is simple binary values and for communication they are converted into ASCII value at presentation layer of OSI model, example character A has ASCII value 65 that can be represented as 8 bit value 01000001 as shown in Fig. 3. Stego key depends on the data-hiding algorithm or stego algorithm as shown in Fig. 2, for example, if we are using simple least significant bit (LSB), then data can be embedded on the LSB bits and for embedding 8 bits we need 8 pixel values.

As very similar to steganography, copyright protection uses digital watermarking or authentication of digital image or document. In watermarking, some small watermark (may be signature or some small portion of data) is embedded into digital media (image or document). Here method of embedding may be the same, but watermark object is very small and consisting only of copyright or authentication information. There are two possible watermarking methods, i.e. visible and non-visible watermarking. In visible type, watermark object is visible and in invisible watermarking, watermark object in invisible as in Fig. 4.

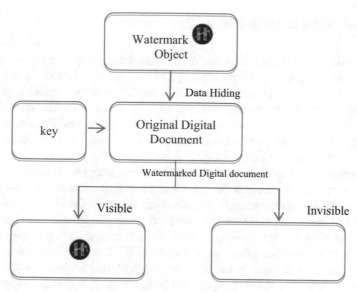

**Fig. 4** Illustration of watermarking process

This paper focuses on digital image as the document and the images may be represented in frequency and spatial domain. In frequency domain, image is denoted in terms of its frequencies; though, in spatial domain, it is denoted by pixel values. In simple terms, multiple frequency bands make the image in frequency domain. Discrete cosine, discrete wavelet, or discrete Fourier transform convert the images to frequency domain. All of them have their characteristics and denotes the images in differently.

Reversible data hiding: Problem associated with data-hiding technique is, when we extract the secret information from stego-image small distortion or noise corrupts the original cover image. In other words, changes occur in the original cover image while accessing the secret information. Solution for that may be some technique which is reversible in nature. So, it can be said that there is one more way to hide secret data through reversible hiding methods which also hides digital images in reversible way, since we know that high volume of secret information are integrated with digital covered images and provide a way to get secret information at receiver end to find actual data, with the help of such data-hiding techniques the receive can extract original information by having such data-hiding reversible technique.

## 2 Illustration of Data-Hiding Technique

Least significant bit (LSB) alteration is an early data-hiding technique. Here, the idea is to utilize image's LSB for storing some secret information. At the time of extraction, LSB bits are scanned in correct order and retrieve the original information.

The LSB methodology embeds a digital message within the LSB plane of a truncated grey scale image. To implement this methodology, browse the truncated grey scale image from the file. Reset the LSB plane of this image to zero. Next, select a message to be written inside the image and transform it into a string of bits. Finally, add the string of bits to the truncated grey scale image and write the watermarked image as a bmp file. To retrieve the data from the watermarked image, first produce a truncated version of this image by transforming its pixel values into even numbers. Subtract the truncated image from its watermarked version to isolate the LSB plane. Then cluster every seven bits along and realize their corresponding American Standard Code for Information Interchange code (ASCII). Use the command window to show your message. It ought to match the message you originally embedded within the image. So information hiding using LSB plane is simple method in spatial domain and also easy to extract.

### 2.1 Survey on Reversible Data-Hiding Technique

In March 2006, Zhicheng Ni et al. gave a technique on reversible data hiding by the use of histogram modification method. Authors presented a novel method that restored the original data which does not have any distortion as of marked image after the original data have been recovered. Main concern with this method is to embed maximum information or increase data embedding rate. As algorithm suggest to use image histogram's either minimum or zero point value of an image and to combine the data into the image after its alteration in pixel grey scale value. Proposed idea is very simple, zero point and peak point is first selected and then each pixel is scanned in sequential order from zero to peak point and embedded the sensitive information. Here sensitive information is also assumed in binary bits format. Authors define zero point as the pixel value which is not assigned and peak point as the highest pixel value point. So in this way by histogram modification method information is embedded into the cover media. Using the same procedure watermarked image can be scanned for extraction and if pixels grey scale value is found $a + 1$ then 1 is extracted otherwise 0. In this way author uses several zero, peak point pair for achieving maximum embedding rate. On various types of images experiments can be applied, which contain common images, texture images, medical images, and all of the 1096 images in the Corel DRAW database, and always got satisfactory results [6].

Chin-Chen Chang proposed a RDH scheme in October 2006 on side match vector quantization (SMVQ) with main concern on the images which are digitally compressed. Authors make some survey that method which is applied for RDH was unable to recover the compressed cover media and data. Any sender trying to mask secret information for compression has to alter its code and the alterations may result in distorting compressed image. After covering the secret information in compression codes, these codes cannot be refurbished or stored for further use. Which means received compression code cannot be used as a carrier after retrieving the masked data. This situation is unproductive for sender and receiver. To make sure RDH author modified the previous image compression technique. In this way, modified SMVQ method is used for preprocessing step and then applies data-hiding technique as accordance to compression and extracts the message efficiently. Here in preprocessing step, cover image is partitioned to some non-overlapping blocks. As first row and column are used for compression code and the data are embedded into the remaining blocks. To create the sub-codebook for every left out block, the upper and left encoded blocks are used. If covert bit is 0, the code word becomes the substance of the stego-image. But if covert bit value is 1, then we look for the code word from the sub-codebook such that the code word is nearest to the code word. Experiments are performed on various $512 \times 512$ standard grey-level images [7].

In September 2007, XiaoTong Wang proposed RDH scheme using difference expansion method. Authors have considered that for hiding information in 2-D vector maps and reversible watermarking is appropriate since the deformation caused by embedding of data can be eliminated after recovering the masked bits. Two types of RDH methods are investigated using difference expansion. In first method, the vertices coordinates are used to cover up the data and hiding is done by altering the differences of the adjoining coordinates. This method attains increased capacity in the maps with extremely correlated coordinates. In place of the coordinates, distances among neighbouring vertices as the cover data are used in the second method. To retrieve the distances from coordinates and the masked data, a set of invertible integer mappings is defined. This method shows better results compared to the earlier one in capacity and invisibility for maps where distances show high correlation. Three types of maps with different features were used during the experiments. The outcomes show that two methods are suitable for dissimilar variety of maps and are firmly reversible [8].

In January 2009 P., Tsai proposed a novel histogram-based RDH for vector quantization compressed images which is some modification of Chin-Chen Chang work proposed in late 2006 [9]. An index-based cover image is generated by prediction vector quantization encoding. Its histogram is explored to embed the secret data by the use of data-hiding procedure based on histogram. Here, authors use SMVQ which is the variation of VQ [10]. For the prediction of current block encoding it uses the relationship between neighbouring blocks. All the image blocks of SMVQ are classified into two types, one is basic block and another is residual block. Basically, the first row and first column blocks are known as basic blocks and the remaining blocks are known as residual blocks. In 2006, Ni et al. [11] suggested a histogram-based method for reversible data hiding that suggest the calculation of all possible pixel values of the cover image to produce the image histogram. Specific pixels are changed to set the secret data. The changed specific pixels can be retrieved during embedded data extraction. Therefore in the proposed work, author focused on encoding using VQ methods so that maximum embedding rate is achieved and also can efficiently recovered the cover image. Embedding strategy is same as the previous proposed by Ni et al. [11].

Yang et al. later in December 2009 proposed histogram-based RDH method by interleaving predictions. Up to this point, many methods for RDH using histogram modification are proposed but their main limitation of this process is it is not relatively predicting actual values that supposed to be compared as a no. of measurable pixels in the process of encoding the image. In this case, one can find that the processes are getting interleaving result by getting the scenario of prediction that causes predictive and pixel values both are measuring with equality. In order to get high peak values for the betterment of the embedding process capacity, one find their values which one predict has been transformed into histogram. That shows the actual values of embedding capacity values are getting heights of peak point values are increased in histogram that one can measure the values are going with the initials of +1 and −1 ranges which is the major objective behind interleaving prediction where one find the pixels ratios between even and odd columns will be predicted simultaneously which is going on similar for even and odd column predictions so that in embedding procedure, odd prediction column values are used for secret data information in histogram so data can be embed easily. Now, even column values will be used to find errors in histogram. The predictive process of finding error of even column will be processed first during the procedure of extracting and reversing. Then, predictive error values of odd columns are processed [12].

Some novel approaches are proposed for efficient image compression and maximum embedding of data up to the point. But a typical way to use image hiding is with some encryption technique. So in early 2011, Xinpeng proposed RDH in encrypted image [13]. With the help of stream cipher technique, one can decompose the text in image format so that the extra information can be embedded with that since the process of modifying a special small portion encrypted data so to process such encrypted to decrypt first need to process it through a key call encryption key

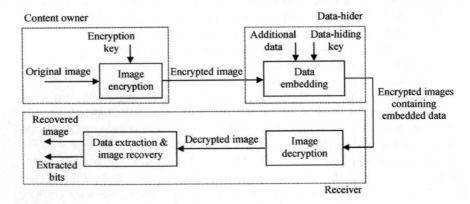

**Fig. 5** RDH in encrypted image

and similar process has been done at other end to get the original data like as shown in Fig. 5.

After this encryption, image is alienated into some non-overlie blocks. These blocks are used to carry one extra bit. For every block, each s2 pixels divide into two sets s1 and s2 by pseudo-randomly. If the extra bit to be embedded is 0, flip the 3 least significant bits (LSB) of every encrypted pixel in set s1 else if extra bit is 1 then flip the 3 LSB of every encrypted pixel in set s2. For extracting data in other end, data hider will first decrypt the image using some XOR operations. The original five most significant bits (MSB) will be retrieved correctly. For a particular pixel, if the embedded bit in the block with the pixel is zero and the pixel belongs to s2, or the embedded bit is 1 and the pixel belongs to s1. Then will check the 3 LSB bits and compare with original image; if they are same, then pixel belongs to s1 else s2 set [13].

In April 2012, Hong [14] proposed a modified version of Zhang's reversible data-hiding method in encrypted images. This idea of Zhang not completely uses the pixels for calculating the smoothness of every block and did not consider the pixel correlations in the border of adjoining blocks. These issues could minimize the accuracy of data extraction. In Wien's new method for calculating the smoothness of blocks reduced the error rate of extracted-bits using side-match scheme.

In February 2013, Xinpeng Zhang proposed a novel RDH method with optimal value transfer which uses the optimal rule of value modification under a payload distortion criterion and is proposed by using an iterative procedure [15]. Embedding method proposed is likely same as histogram shifting or difference expansion method but they are additionally generating optimum matrix as shown in Fig. 6.

As shown in Fig. 6, authors additionally generate the optimal transfer matrix for pure reversibility and maximize payload data.

A 2-D difference-histogram modification method [16] for RDH was suggested by Xiaolong Li et al. in July 2013. In this method initially, by taking into account each pixel-pair and its perspective, a sequence comprising of pairs of difference values is found. After that, a 2-D difference-histogram is produced by including

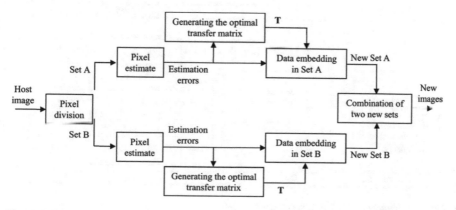

**Fig. 6** RDH proposed by Xinpeng

the frequency of the resultant difference-pairs. At last, reversible data embedding is implemented according to a particularly designed difference-pair-mapping (DPM). Here, the difference-pair-mapping is an injective mapping defined on difference-pairs. It is simply a natural extension of expansion embedding and shifting techniques. Actually this work is slightly the modification of Lee et al.'s method which modifies or embeds data in one direction and achieves reversibility. In this, up and down are

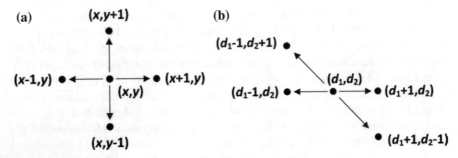

**Fig. 7** **a** Modifying either $x$ or $y$ by 1, $(x, y)$ has four modification directions. **b** The corresponding difference-pair $(d_1, d_2)$ also has four modification directions, where $d_1 = (x - y)$, $d_2 = (y - z)$, and $z$ is a prediction of $y$

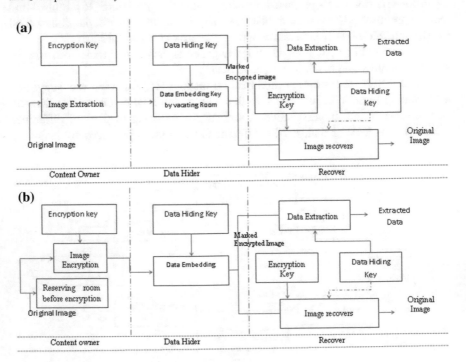

**Fig. 8** **a** Framework VRAE. **b** Framework RRBE

proposed as two modification directions, and are permitted in data embedding as shown in Fig. 7.

In March 2013, Kede Ma et al. proposed a new concept of reserving room before encryption [17]. Authors predicted that reserving room before encryption is more efficient than traditional vacating room after encryption method. In the proposed work, author did the same work of histogram modification as before, this can be illustrated as in Fig. 8.

# 3 Comparison and Discussion

After making a brief survey of all possible aspects of RDH, four important areas can be concluded. So in this section, we complete the comparison of RDH using data compression technique, difference expansion, histogram shifting methods and last error prediction method.

In Difference expansion method, secret data is embedded in the difference of pixel value.

For a couple of pixel value $(x, y)$ in a grey scale image, $x, y \in Z, 0 \le x, y \ge 255$, defined there integer average $l$ and difference $h$ as:

$$l = \left| \frac{x + y}{2} \right|, \quad h = x - y \tag{1}$$

where $|.|$ is used as floor function. The inverse of transform 1 is

$$x = l + \left| \frac{h + 1}{2} \right|, \quad y = l - \left| \frac{h}{2} \right| \tag{2}$$

As grey scale value is bounded to 0–255, we can show

$$|h| \le \min(2(255 - l), 2 * l + 1) \tag{3}$$

Equation 3 can be used for preventing overflow and underflow conditions. A simple scheme for data embedding may be by using least significant bits only.

In histogram-based method, sensitive information hides in the difference of histogram, but we cannot embed information if image having a linear histograms.

Below this, we present data embedding by these four methods, discussed earlier. Experiments are performed on some well-known images having size $512 \times 512$.

| Embedded rate | | 0.1 | 0.2 | 0.3 | 0.4 | 0.5 |
|---|---|---|---|---|---|---|
| Lena | Histogram-based method | 54 | 50.8 | 47.76 | 45.65 | 43.53 |
| | Difference expansion method | 44.20 | 42.86 | 41.55 | 37.66 | 36.15 |
| | Reserving room before encryption | 52.33 | 49.07 | 45.00 | 40.65 | 35.84 |
| Barbara | Histogram-based method | 54.37 | 49.435 | 46.276 | 42.706 | 40.540 |
| | Difference expansion method | 44.20 | 42.86 | 41.55 | 37.66 | 36.15 |
| | Reserving room before encryption | 52.461 | 47.68 | 43.46 | 39.24 | 25.92 |
| Cameraman | Histogram-based method | 48.94 | 44.42 | 39.469 | 35.552 | 33.16 |
| | difference expansion method | 44.20 | 42.86 | 41.55 | 37.66 | 36.15 |
| | Reserving room before encryption | 51.63 | 50.49 | 48.24 | 46.21 | 43.19 |
| Baboon | Histogram-based method | 46.94 | 41.42 | 37.469 | 34.552 | 32.16 |
| | Difference expansion method | 44.20 | 42.86 | 41.55 | 37.66 | 36.15 |
| | Reserving room before encryption | 46.17 | 40.68 | 35.87 | 31.16 | 25.92 |
| Peppers | Histogram-based method | 52.127 | 47.019 | 44.210 | 41.762 | 39.905 |
| | Difference expansion method | 44.20 | 42.86 | 41.55 | 37.66 | 36.15 |
| | Reserving room before encryption | 51.02 | 46.00 | 42.08 | 36.91 | 27.09 |

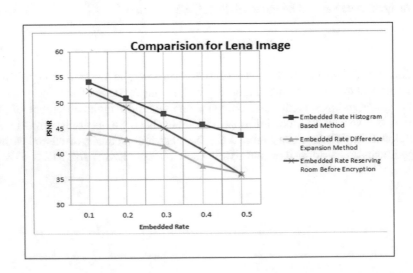

**Fig. 9** Classification of reversible data hiding

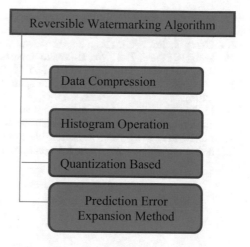

## 4  Conclusion

So as per our literature survey, it is very clear that there are four dimensions for reversible data hiding namely compression-based method, histogram modification-based method, quantization-based method and last expansion-based method as shown in Fig. 9.

## References

1. Khan, A., et al.: A recent survey of reversible watermarking techniques. Inf. Sci. (2014)
2. Cox, I.J., Miller, M.L., Bloom, J.A., Fridrich, J., Kalker, T.: Digital Watermarking and Steganography, 2nd edn. Morgan Kaufmann (2008)
3. EL-Emam, N.N.: Hiding a large amount of data with high security using steganography algorithm. J. Comput. Sci. **3**(4), 223–232 (2007)
4. Petitcolas, F.A.P.: Information hiding—a survey. Proc. IEEE (Special Issue on Protection of Multimedia Content) **87**(7), 1062–1078 (1999)
5. Halpern J.Y.: Anonymity and information hiding in multiagent systems. In: 16th IEEE Computer Security Foundations Workshop, 2003. Proceedings, pp. 75–88
6. Ni, Z., Shi, Y.-Q., Ansari, N., Su, W.: Reversible data hiding. IEEE Trans. Circuits Syst. Video Technol. **16**(3) (2006)
7. Chang, C.-C.: A reversible data hiding scheme based on side match vector quantization. IEEE Trans. Circuits Syst. Video Technol. **16**(10) (2006)
8. Wang, X.T., Shao, C.Y., Xu, X.G., Niu, X.M.: Reversible data-hiding scheme for 2-D vector maps based on difference expansion. IEEE Trans. Inf. Forensics Secur. **2**(3) (2007)
9. Tsai, P.: Histogram-based reversible data hiding for vector quantization-compressed images. IET Image Process. (2009)
10. Gray, R.M.: Vector quantization. IEEE ASSP Mag. **1**(2), 4–49 (1984)
11. Ni, Z., Shi, Y.Q., Ansari, N., Su, W.: Reversible data hiding. IEEE Trans. Circuits Syst. Video Technol. **16**(3), 354–361 (2006)

12. Yang, C.-H., Tsai, M.-H.: Improving histogram-based reversible data hiding by interleaving predictions. IET Image Process. (2009)
13. Zhang, X.: Reversible data hiding in encrypted image. IEEE Sig. Process. Lett. **18**(4) (2011)
14. Hong, W., Chen, T.-S., Wu, H.-Y.: An improved reversible data hiding in encrypted images using side match. IEEE Sig. Process. Lett. **19**(4) (2012)
15. Zhang, X.: Reversible data hiding with optimal value transfer. IEEE Trans. Multimed. **15**(2) (2013)
16. Li, X., Zhang, W., Gui, X., Yang, B.: A novel reversible data hiding scheme based on two-dimensional difference-histogram modification. IEEE Trans. Inf. Forensics Secur. **8**(7) (2013)
17. Ma, K., Zhang, W., Zhao, X.: Reversible data hiding in encrypted images by reserving room before encryption. IEEE Trans. Inf. Forensics Secur. **8**(3) (2013)

# Investigation of Feature Selection Techniques on Performance of Automatic Text Categorization

**Dilip Singh Sisodia and Ankit Shukla**

## 1 Introduction

Nowadays, digital documentation is increasing at a very fast pace, and it is very important to maintain the classification of digital documents. The main aim of digital document classification is to categorize the documents into predefined classes. It is an active research area for the information retrieval [1] and machine learning from the digital text documentation. There are many supervised algorithms which are employed on the digital text documents for the classification such as support vector machine [2], Naïve Bayes [3], decision tree [4], and nearest neighbors [5].

There are two phases of text categorization [6] of digital documents: One is the training phase, and the second is classification testing phase. Earlier, subject indexing and feature extraction method [7] were used for text categorization. However, these methods are not very much successful for the classification. Text categorization methods are based on the term frequency and inverted term frequency and count the frequencies of the term but not consider the position of the term. Therefore, these methods were not efficient in articulating the class for the text data. In each data, the position of the term is very relevant for the identification of the documents.

The remaining paper is organized as follows: Sect. 2 discusses the related work. In Sect. 3, material and methodology used for this work are discussed. Section 4 describes the experimental results and discussions. Lastly, Sect. 5 concludes this study.

D. S. Sisodia (✉)
National Institute of Technology, Raipur, India
e-mail: dssisodia.cs@nitrr.ac.in

A. Shukla
Jaypee University of Engineering & Technology, Guna, India
e-mail: ankeetshk@gmail.com

© Springer Nature Singapore Pte Ltd. 2019
R. K. Shukla et al. (eds.), *Data, Engineering and Applications*,
https://doi.org/10.1007/978-981-13-6347-4_7

## 2  Related Work

Earlier, the text classification was done manually, but those classifications were not at all efficient. After that, many classification schemes came to existence such as subject indexing [8], term frequency [9], Gini index [10], mutual information, and information gain [11]. Till now, a significant amount of research has been done in automatic text categorization (ATC). Term frequency and subject indexing also used for classification, but these techniques were using the phenomenon of term redundancy [12] and subject index but missing the relevancy of the term. Gini index is also a global feature selection method for text classification. It is an improved version attribute selection algorithm. Currently, the weighted feature selection [13] algorithms are used for automatic text categorization since it is based on the mutual information [14, 15] of the term of the dataset. Mutual information and maximum entropy classification [16] are the basic techniques which are used by the researcher for machine learning and information retrieval from the text document.

## 3  Material and Methodology

### 3.1  Data Source

Four datasets have been taken from the Knowledge Extraction based on Evolutionary Learning (KEEL) repository text classification datasets. It contains preprocessed data of text document of Ohsumed test collection which is a subset of the MEDLINE database. The MEDLINE database is a collection of the bibliographic database of important, peer-reviewed medical literature maintained by the National Library of Medicine. Ohsumed test classification [17] is the collection of each dataset which contains 100 attributes which are enough to test various feature selection algorithms. The brief description of the used dataset is given in Table 1.

### 3.2  Methodology

Before doing any classification, we need to do preprocessing of dataset. Since dataset is very large and has an enormous number of the attribute, we have to reduce the number of the attribute in the dataset using preprocessing step known as feature selection. There are various feature selection algorithms available, but we will use only those feature selection algorithms which are available in the feature selection toolbox developed at UTIA of the Czech Academy of Sciences. The methodology works as shown in Fig. 1.

**Table 1** A brief description of used data sets

| Dataset name | Number of instances | Class labels | Number of features |
|---|---|---|---|
| 6 Ketoprostaglandin F1 $\alpha$ | 1003 | negative 6-ketoprostaglandin F1 $\alpha$, positive 6-ketoprostaglandin F1 $\alpha$, | 100 |
| Brain chemistry data | 1003 | negative brain chemistry positive brain chemistry | 100 |
| Heart valve data | 1003 | negative heart valve positive heart valve | 100 |
| Uric acid data | 1003 | negative uric acid positive uric acid | 100 |

**Feature Selection Algorithms** *CMIM*. The conditional mutual information maximization (CMIM) [18] algorithm selects a subset of a feature from dataset to minimize the number of features. The selected features carry more relevant information of data according to mutual information and save computational time.

*mRMR*. The minimum redundancy maximum relevance (mRMR) [18] select features are having less redundant data to minimize feature redundancy and high correlation to maximize feature relevance. The two usually used objective functions in mRMR are mutual information difference criterion (MID) and mutual information quotient criterion (MIQ).

*JMI*. The joint mutual information (JMI) [19] uses information theory to calculate the mutual information and entropy between any random variables together for feature selection. The representation is shown in Eq. (1).

$$I(x, \ y) = H(x) - H\,(x|y) \tag{1}$$

where $I$ is mutual information and $H$ is entropy.

*Condred*. In the condition redundancy [20] feature selection method, the race condition is overcome. The race condition is occurred due to the redundancy of term which is not related to the classification of a text document to predefined classes and statistical property.

*MIFS*. Mutual information feature selection (MIFS) [21] algorithm is entirely based on the mutual information computed for each term of the dataset. Mutual information of dataset is more reliable data as compared to the frequency of data for the classification of a text document. MIFS gives a more precise result, but it is a little bit slower since it calculates the weight of each data and then weight frequency.

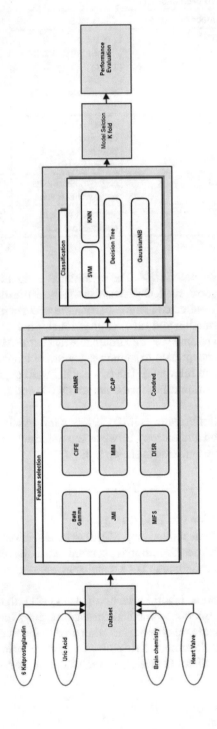

**Fig. 1** Methodology of text classification via feature selection

*ICAP*. In Interaction capping (ICAP) [22] feature selection algorithm features are sorted using the interaction of their term with other term using the information capping.

*DISR*. The double-input symmetrical relevance (DISR) [23] feature selection algorithm combines two main properties of variable complementarily, and the collection of the feature gives a different result. The most promising set is $d - 1$ if there is no information about the relation of the variable in datasets.

*CIFE*. The conditional infomax feature extraction (CIFE) [24] algorithm is based on information theory. In this feature selection, the systematical study of the structure of the document is done. It improves the performance of joint-class relevant detail by reducing class redundancy of dataset [8].

*BetaGamma*. The BetaGamma [25] is conditional mutual information-based feature selection algorithm. In this algorithm, beta and gamma are two values that maintain the weight of a feature by their relevance. Normally, the value of $\beta$ (beta) and $\gamma$ (gamma) is zero.

**Classification Algorithm** *Support Vector Machine*. SVM is a supervised learning technique and segregates classes using hyperplane for the classification using feature values of $N$ instances.

*Decision Tree*. A decision tree is a predictive model used for the classification based on the tree model. In this supervised algorithm, the datasets are broken down into a subset and create an association with that subset in a form decision tree having the node as decision node, intermediate node, and leave node.

*K-nearest neighbors*. In this method, the prediction function depends on the approximated locality, and Euclidean distance [25], Chebyshev norm, or Mahalanobis distance is used for distance computation.

*Gaussian Naïve Bayes*. This algorithm is based on the probabilistic classifier and relies on the well-known theorem—Bayes' theorem. It is also very popular for text categorization method. In this algorithm, the following formula in Eq. (2) are implied

$$P(c/X) = \sum_{n=1}^{\infty} P(xi/c) \qquad (2)$$

where $P(c/X)$ is the posterior probability.

# 4  Experimental Results and Discussions

This section summarizes the simulation result performed on four text categorical data. We consider four classifiers and nine feature selection technique for the sake of performance evaluation. Results are annotated for each classifier and feature selection technique pair. The accuracy values are recorded and listed in Tables 2, 3, 4, and 5. The feature selection techniques referred in this study are from the filter-based approach, which requires the number of the feature as an input parameter. Due to

**Table 2** Impact of feature selection algorithm on 6-ketoprostaglandin F1 $\alpha$

| Classifiers | Feature selection technique | Number of features | | | | |
|---|---|---|---|---|---|---|
| | | 10 | 20 | 30 | 40 | 50 |
| SVM | BetaGamma | 0.996 | 0.980 | 0.988 | 0.972 | 0.980 |
| | CIFE | 0.976 | 0.980 | 0.988 | 0.984 | 0.984 |
| | CMIM | 0.992 | 1 | 0.988 | 0.992 | 0.992 |
| | Condred | 0.988 | 1 | 0.984 | 0.992 | 0.988 |
| | DISR | 1 | 0.984 | 0.984 | 0.996 | 0.996 |
| | ICAP | 0.988 | 0.972 | 0.984 | 0.980 | 0.988 |
| | JMI | 0.976 | 0.992 | 0.996 | 0.992 | 0.988 |
| | MIFS | 0.984 | 0.992 | 0.984 | 0.996 | 0.992 |
| | MRMR | 0.984 | 1 | 0.996 | 0.996 | 0.996 |
| KNN | BetaGamma | 0.964 | 0.976 | 0.972 | 0.988 | 0.964 |
| | CIFE | 0.988 | 0.976 | 0.968 | 0.972 | 0.980 |
| | CMIM | 0.996 | 0.984 | 0.972 | 0.964 | 0.964 |
| | Condred | 0.976 | 0.980 | 0.988 | 0.980 | 0.980 |
| | DISR | 0.980 | 0.992 | 0.972 | 0.988 | 0.980 |
| | ICAP | 0.980 | 0.976 | 0.968 | 0.996 | 0.976 |
| | JMI | 0.996 | 0.980 | 0.984 | 0.972 | 0.972 |
| | MIFS | 0.996 | 0.976 | 0.988 | 0.976 | 0.960 |
| | MRMR | 0.984 | 0.988 | 0.996 | 0.964 | 0.976 |
| DT | BetaGamma | 0.976 | 0.980 | 0.972 | 0.956 | 0.972 |
| | CIFE | 0.984 | 0.992 | 0.972 | 0.972 | 0.940 |
| | CMIM | 0.988 | 0.976 | 0.984 | 0.984 | 0.984 |
| | Condred | 0.984 | 0.984 | 0.988 | 0.972 | 0.980 |
| | DISR | 0.980 | 0.984 | 0.984 | 0.984 | 0.988 |
| | ICAP | 0.984 | 0.992 | 0.968 | 0.980 | 0.972 |
| | JMI | 0.988 | 0.984 | 0.976 | 0.980 | 0.968 |
| | MIFS | 0.992 | 0.984 | 0.988 | 0.972 | 0.968 |
| | MRMR | 0.976 | 0.972 | 0.984 | 0.972 | 0.964 |
| GaussianNB | BetaGamma | 0.984 | 0.984 | 0.964 | 0.972 | 0.960 |
| | CIFE | 1 | 0.992 | 0.984 | 0.960 | 0.972 |
| | CMIM | 0.996 | 0.980 | 0.992 | 0.976 | 0.984 |
| | Condred | 0.964 | 0.908 | 0.940 | 0.928 | 0.920 |
| | DISR | 0.988 | 1 | 0.984 | 0.984 | 0.980 |
| | ICAP | 0.976 | 0.996 | 0.984 | 0.984 | 0.992 |
| | JMI | 0.984 | 0.980 | 0.952 | 0.964 | 0.952 |
| | MIFS | 0.992 | 0.988 | 0.984 | 0.964 | 0.964 |
| | MRMR | 0.992 | 0.984 | 0.972 | 0.988 | 0.976 |

**Table 3** Impact of feature selection algorithm on uric acid data

| Classifiers | Feature selection technique | Number of features | | | | |
|---|---|---|---|---|---|---|
| | | 10 | 20 | 30 | 40 | 50 |
| SVM | BetaGamma | 0.996 | 0.992 | 0.984 | 0.984 | 0.984 |
| | CIFE | 0.996 | 0.980 | 0.992 | 0.992 | 0.988 |
| | CMIM | 0.980 | 0.996 | 0.996 | 0.988 | 1 |
| | Condred | 0.972 | 0.968 | 0.984 | 0.980 | 0.988 |
| | DISR | 0.992 | 0.992 | 0.984 | 0.992 | 0.988 |
| | ICAP | 0.984 | 0.988 | 0.988 | 0.992 | 0.968 |
| | JMI | 0.992 | 0.988 | 0.972 | 0.984 | 0.988 |
| | MIFS | 0.980 | 0.988 | 0.980 | 0.980 | 0.984 |
| | MRMR | 0.984 | 0.992 | 0.996 | 0.992 | 0.988 |
| KNN | BetaGamma | 0.976 | 0.984 | 0.964 | 0.964 | 0.964 |
| | CIFE | 0.992 | 0.968 | 0.952 | 0.960 | 0.956 |
| | CMIM | 0.984 | 0.944 | 0.948 | 0.948 | 0.956 |
| | Condred | 0.992 | 0.960 | 0.960 | 0.960 | 0.984 |
| | DISR | 0.964 | 0.968 | 0.936 | 0.948 | 0.960 |
| | ICAP | 0.980 | 0.980 | 0.972 | 0.976 | 0.940 |
| | JMI | 0.984 | 0.976 | 0.984 | 0.976 | 0.952 |
| | MIFS | 0.988 | 0.960 | 0.964 | 0.964 | 0.964 |
| | MRMR | 0.976 | 0.972 | 0.964 | 0.960 | 0.964 |
| DT | BetaGamma | 0.960 | 0.944 | 0.976 | 0.944 | 0.968 |
| | CIFE | 0.984 | 0.960 | 0.968 | 0.960 | 0.932 |
| | CMIM | 0.996 | 0.976 | 0.976 | 0.964 | 0.960 |
| | Condred | 0.980 | 0.980 | 0.964 | 0.952 | 0.956 |
| | DISR | 0.968 | 0.972 | 0.968 | 0.988 | 0.964 |
| | ICAP | 0.988 | 0.964 | 0.972 | 0.968 | 0.932 |
| | JMI | 0.980 | 0.976 | 0.964 | 0.968 | 0.956 |
| | MIFS | 0.980 | 0.960 | 0.980 | 0.960 | 0.956 |
| | MRMR | 0.988 | 0.980 | 0.968 | 0.968 | 0.972 |
| GaussianNB | BetaGamma | 0.992 | 0.952 | 0.976 | 0.936 | 0.912 |
| | CIFE | 1 | 0.972 | 0.952 | 0.944 | 0.928 |
| | CMIM | 0.992 | 0.988 | 0.992 | 0.972 | 0.980 |
| | Condred | 0.972 | 0.972 | 0.972 | 0.960 | 0.972 |
| | DISR | 0.984 | 0.988 | 0.980 | 0.980 | 0.960 |
| | ICAP | 0.996 | 0.980 | 0.964 | 0.956 | 0.928 |
| | JMI | 0.980 | 0.972 | 0.980 | 0.988 | 0.960 |
| | MIFS | 0.972 | 0.992 | 0.984 | 0.988 | 0.984 |
| | MRMR | 0.980 | 0.976 | 0.996 | 0.980 | 0.984 |

**Table 4** Impact of feature selection algorithm on heart valve data

| Classifiers | Feature selection technique | Number of features | | | | |
|---|---|---|---|---|---|---|
| | | 10 | 20 | 30 | 40 | 50 |
| SVM | BetaGamma | 0.984 | 0.964 | 0.988 | 0.976 | 0.992 |
| | CIFE | 0.972 | 0.988 | 0.976 | 0.984 | 0.976 |
| | CMIM | 0.976 | 0.976 | 0.980 | 0.968 | 0.976 |
| | Condred | 0.960 | 0.944 | 0.980 | 0.988 | 0.980 |
| | DISR | 0.968 | 0.972 | 0.984 | 0.968 | 0.980 |
| | ICAP | 0.976 | 0.968 | 0.976 | 0.992 | 0.961 |
| | JMI | 0.956 | 0.976 | 0.980 | 0.980 | 0.988 |
| | MIFS | 0.980 | 0.992 | 0.980 | 0.968 | 0.968 |
| | MRMR | 0.980 | 0.996 | 0.984 | 0.976 | 0.988 |
| KNN | BetaGamma | 0.964 | 0.992 | 0.940 | 0.960 | 0.944 |
| | CIFE | 0.968 | 0.968 | 0.956 | 0.952 | 0.936 |
| | CMIM | 0.972 | 0.976 | 0.952 | 0.952 | 0.940 |
| | Condred | 0.952 | 0.976 | 0.960 | 0.968 | 0.960 |
| | DISR | 0.968 | 0.972 | 0.968 | 0.928 | 0.968 |
| | ICAP | 0.972 | 0.968 | 0.984 | 0.960 | 0.960 |
| | JMI | 0.980 | 0.976 | 0.972 | 0.948 | 0.964 |
| | MIFS | 0.980 | 0.960 | 0.972 | 0.956 | 0.976 |
| | MRMR | 0.980 | 0.968 | 0.960 | 0.976 | 0.948 |
| DT | BetaGamma | 0.968 | 0.992 | 0.952 | 0.964 | 0.952 |
| | CIFE | 0.968 | 0.960 | 0.980 | 0.968 | 0.960 |
| | CMIM | 0.980 | 0.964 | 0.980 | 0.964 | 0.940 |
| | Condred | 0.940 | 0.952 | 0.952 | 0.952 | 0.972 |
| | DISR | 0.956 | 0.992 | 0.968 | 0.944 | 0.940 |
| | ICAP | 0.984 | 0.980 | 0.960 | 0.976 | 0.936 |
| | JMI | 0.960 | 0.952 | 0.960 | 0.940 | 0.928 |
| | MIFS | **0.992** | 0.964 | 0.956 | 0.964 | 0.952 |
| | MRMR | 0.968 | 0.968 | 0.968 | 0.968 | 0.964 |
| GaussianNB | BetaGamma | 0.956 | 0.940 | 0.936 | 0.948 | 0.948 |
| | CIFE | 0.972 | 0.944 | 0.956 | 0.936 | 0.920 |
| | CMIM | 0.940 | 0.976 | 0.976 | 0.976 | 0.956 |
| | Condred | 0.948 | 0.928 | 0.920 | 0.940 | 0.944 |
| | DISR | 0.948 | 0.972 | 0.968 | 0.976 | 0.972 |
| | ICAP | 0.976 | 0.968 | 0.968 | 0.960 | 0.944 |
| | JMI | 0.972 | 0.972 | 0.972 | 0.964 | 0.960 |
| | MIFS | 0.972 | 0.972 | 0.956 | 0.956 | 0.964 |
| | MRMR | 0.960 | 0.968 | 0.956 | 0.964 | 0.956 |

**Table 5** Impact of feature selection algorithm on brain chemistry data

| Classifiers | Feature selection technique | Number of features | | | | |
|---|---|---|---|---|---|---|
| | | 10 | 20 | 30 | 40 | 50 |
| SVM | BetaGamma | 0.992 | 0.992 | 0.988 | 0.980 | 0.988 |
| | CIFE | 0.992 | 0.988 | 0.988 | 0.984 | 0.992 |
| | CMIM | 0.992 | 0.992 | 0.996 | 0.980 | 0.988 |
| | Condred | 0.984 | 0.980 | 0.980 | 0.968 | 0.968 |
| | DISR | 0.992 | 0.984 | 0.976 | 0.964 | 0.992 |
| | ICAP | 0.984 | 0.976 | 0.952 | 0.976 | 0.948 |
| | JMI | 0.976 | 0.984 | 0.960 | 0.976 | 0.992 |
| | MIFS | 0.996 | 0.992 | 0.988 | 0.958 | 0.980 |
| | MRMR | 0.980 | 0.980 | 0.992 | 0.944 | 1 |
| KNN | BetaGamma | 0.976 | 0.952 | 0.972 | 0.964 | 0.964 |
| | CIFE | 0.964 | 0.968 | 0.944 | 0.932 | 0.952 |
| | CMIM | 0.960 | 0.956 | 0.964 | 0.948 | 0.956 |
| | Condred | 0.988 | 0.980 | 0.968 | 0.976 | 0.968 |
| | DISR | 0.972 | 0.980 | 0.984 | 0.984 | 0.964 |
| | ICAP | 0.972 | 0.964 | 0.968 | 0.976 | 0.948 |
| | JMI | 0.992 | 0.972 | 0.980 | 0.956 | 0.944 |
| | MIFS | 0.980 | 0.980 | 0.980 | 1 | 0.936 |
| | MRMR | 0.988 | 0.976 | 0.992 | 0958 | 0.936 |
| DT | BetaGamma | 0.984 | 0.968 | 0.956 | 0.956 | 0.924 |
| | CIFE | 0.988 | 0.980 | 0.952 | 0.952 | 0.944 |
| | CMIM | 0.992 | 0.992 | 0.952 | 0.956 | 0.964 |
| | Condred | 0.992 | 0.968 | 0.964 | 0.956 | 0.988 |
| | DISR | 0.984 | 0.956 | 0.976 | 0.984 | 0.980 |
| | ICAP | 0.992 | 0.992 | 0.992 | 0.924 | 0.924 |
| | JMI | 0.980 | 0.948 | 0.972 | 0.980 | 0.956 |
| | MIFS | 0.976 | 0.980 | 0.972 | 0.988 | 0.980 |
| | MRMR | 0.980 | 0.960 | 0.980 | 0.980 | 0.972 |
| GaussianNB | BetaGamma | 0.996 | 0.988 | 0.972 | 0.932 | 0.908 |
| | CIFE | 0.992 | 0.980 | 0.964 | 0.932 | 0.924 |
| | CMIM | 1 | 0.992 | 0.992 | 0.992 | 0.992 |
| | Condred | 0.964 | 0.948 | 0.928 | 0.968 | 0.968 |
| | DISR | 0.984 | 0.968 | 0.980 | 0.956 | 0.988 |
| | ICAP | 0.968 | 0.984 | 0.992 | 0.992 | 0.984 |
| | JMI | 0.992 | 0.984 | 0.992 | 0.980 | 0.984 |
| | MIFS | 0.972 | 0.996 | 0.980 | 0.980 | 0.972 |
| | MRMR | 0.984 | 0.936 | 0.988 | 0.976 | 0.968 |

uncertainty in the selection of optimal features, we performed multiple experiments by initializing the ten features with an interval of 10. In total, we capture the five instances in multiple of ten features.

Table 2 lists the experimental result performed on ketoprostaglandin dataset. The accuracy values in the table indicate the highest value when support vector machine is aligned with DISR on ten features. When selected features are 20, then the three feature selection techniques, CMIM, conditional reduced (Condred), and MRMR, produce 100% accuracy.

The performance on uric acid data is noted in Table 3. The table reveals that the CIFE feature selection techniques with Naïve Bayesian classifiers achieved 100% accuracy with ten features only.

Decision tree performance in heart valve data is delivering the maximum accuracy. MIFS feature selection technique with ten features achieves 99.2% of accuracy. Experimental results are listed in Table 4.

In brain chemistry data, the combination of Naïve Bayes algorithm along with CMIM feature selection technique has the highest classification rate. This pair of classifier feature selection achieves 100% accuracy when the number of the feature is selected as 10. Table 5 reports the experimental outcome.

The objectives of these experiments were to extract out the best feature selection and classifier combination so that the choice of making an efficient model could be effortless. However, there is no single pair that can be identified, but still, the use of Naïve Bayes classifier along with CMIM feature selection technique could be an optimal choice for text categorization model.

# 5 Conclusion

In this paper, nine weighted feature selection algorithms are used in the four-text classification preprocessed dataset from KEEL repository. The feature selection is performed with the different number of features ranging from 10 to 50 at the interval of 10 features. This experiment shows improvement in the performance of classification for text documentation categorization on using weighted feature selection. The experimental results concluded that mutual information based feature selection algorithm improves the result of text classification significantly. The weighted feature selection methods are also work well because of the relevance of position of the term used in the text document, and it also reduces factor of redundancy while classification of documents.

# References

1. Joachims, T.: Text categorization with support vector machines: learning with many relevant features. In: Proceedings of the 10th European Conference on Machine Learning ECML '98, pp. 137–142 (1998)
2. Markowetz, F.: Classification by support vector machines. In: Discrete Methods in Epidemiology, pp. 1–9 (2000)
3. Leung, K.M.: Naive Bayesian Classifier (2007)
4. Friedl, M.A., Brodley, C.E.: Decision tree classification of land cover from remotely sensed data. Remote Sens. Environ. **61**, 399–409 (1997)
5. Cai, Y., Ji, D., Cai, D.: A KNN research paper classification method based on shared nearest neighbor. In: Proceedings of the 8th NTCIR Workshop Meeting on Evaluation of Information Access Technologies: Information Retrieval, Question Answering and Cross-Lingual Information Access, pp. 336–340 (2010)
6. Ladha, L., Deepa, T.: Feature selection methods and algorithms. Int. J. Comput. Sci. Eng. **3**, 1787–1797 (2011)
7. Brown, G., Pocock, A., Zhao, M.-J., Lujan, M.: Conditional likelihood maximisation: a unifying framework for mutual information feature selection. J. Mach. Learn. Res. **13**, 27–66 (2012)
8. Albrechtsen, H.: Subject analysis and indexing. From automated indexing to domain analysis. Indexer **18**, 219–224 (1993)
9. Amati, G., van Rijsbergen, C.J.: Term frequency normalization via Pareto distributions. Adv. Inf. Retr. **2291**, 183–192 (2002)
10. Gini coefficient
11. Kraskov, A., Stögbauer, H., Grassberger, P.: Estimating mutual information. Phys. Rev. E-Stat. Nonlinear Soft Matter Phys. **69** (2004)
12. Boulis, C., Ostendorf, M.: Text classification by augmenting the bag-of-words representation with redundancy compensated bigrams. In: Workshop on Feature Selection in Data Mining, pp. 9–16 (2005)
13. Agre, G., Dzhondzhorov, A.: A weighted feature selection method for instance-based classification. In: International Conference on Artificial Intelligence: Methodology, Systems, and Applications, pp. 14–25 (2016)
14. Pluim, J.P.W., Maintz, J.B.A.A., Viergever, M.A.: Mutual-Information-Based Registration of Medical Images: A survey (2003)
15. Li, W.: Mutual information functions versus correlation functions. J. Stat. Phys. **60**, 823–837 (1990)
16. Nigam, K., Lafferty, J., Mccallum, A.: Using maximum entropy for text classification. In: IJCAI-99 Workshop on Machine Learning for Information Filtering, vol. 1, pp. 61–67 (1999)
17. Alcalá-Fdez, J., Fernández, A., Luengo, J., Derrac, J., García, S., Sánchez, L., Herrera, F.: KEEL data-mining software tool: data set repository, integration of algorithms and experimental analysis framework. J. Mult. Valued Log. Soft Comput. **17**, 255–287 (2011)
18. Fleuret, F.: Fast binary feature selection with conditional mutual information. J. Mach. Learn. Res. **5**, 1531–1555 (2004)
19. Bennasar, M., Hicks, Y., Setchi, R.: Feature selection using joint mutual information maximisation. Expert Syst. Appl. **42**, 8520–8532 (2015)
20. Long, W.C., Swiney, K.M., Harris, C., Page, H.N., Foy, R.J.: Effects of ocean acidification on juvenile red king crab (*Paralithodes camtschaticus*) and tanner crab (*Chionoecetes bairdi*) growth, condition, calcification, and survival. PLoS ONE **8** (2013)
21. Rades, M., Ewins, D.: Mifs and macs in modal analysis. In: Modal Analysis Conference (IMAC-20), pp. 771–778 (2002)
22. Jakulin, A.: Machine learning based on attribute interactions. PhD thesis, pp. 1–252 (2005)
23. Bar-Nun, A., Dimitrov, V., Tomasko, M.: Titan's aerosols: comparison between our model and DISR findings. Planet. Space Sci. **56**, 708–714 (2008)

24. Fischer, M., Stone, M., Liston, K., Kunz, J., Singhal, V.: Multi-stakeholder collaboration : the CIFE iRoom. In: International Council for Research and Innovation in Building and Construction. CIB W78 Conference, pp. 12–14 (2002)
25. Lewis, D.: Feature selection and feature extract ion for text categorization. In: Speech and Natural Language: Proceedings of a Workshop Held at Harriman, New York, 23–26 Feb 1992

# Identification and Analysis of Future User Interactions Using Some Link Prediction Methods in Social Networks

**Krishna Das and Smriti Kumar Sinha**

# 1 Introduction

Social networks, for example, Facebook, Twitter and so on have prodded heaps of research in connect forecast and recommendation, which go for foreseeing in secret or missing associations in light of existing structure in a network. Connections in a social network are spoken to by an arrangement of nodes and edges, in which nodes are on-screen characters, edges are associations between those on-screen characters. In a true setting, edge data is absent because of various reasons, for example, deficient information accumulation processor, on the other hand, instability of connections or asset constraints. Moreover, to foresee future associations in a dynamic system is additionally an interesting issue. Person to person communication sites might want to tweak new companion proposals for clients, intelligence offices can counteract and foresee criminal exercises by checking potential connections in malevolent systems, financial associations might want to recognize deceitful activities by assessing value-based systems and so on. Consequently, setting up a strong machine learning model to effectively anticipate potential connections are beneficial.

In this report, we investigated an accumulation of cutting edge approaches in connect forecast region, roused by which, we conducted tests utilizing different techniques. More or less, we have a trained data le organized as adjoining list and, a test data le organized as source-goal datasets. For each match in the test, we need to decide if a given edge is genuine or not in the adjacent list data le. Naturally, the training network is a subgraph acquired from the entire Twitter data. Fundamentally, we utilized both supervised and unsupervised techniques to our learning models and

K. Das (✉) · S. K. Sinha
Department of Computer Science and Engineering,
Tezpur University, Tezpur 784028, Assam, India
e-mail: krishnadas.mca@gmail.com

S. K. Sinha
e-mail: smritiKumarSinha@gmail.com

© Springer Nature Singapore Pte Ltd. 2019
R. K. Shukla et al. (eds.), *Data, Engineering and Applications*,
https://doi.org/10.1007/978-981-13-6347-4_8

shockingly found that the unsupervised strategy outperformed supervised strategies, which is additionally examined in the following segments.


## 2 Related Work

The connection expectation issues are complex issues because of its sparse property. Research has normally handled this by utilizing unsupervised methodologies, and a large portion of which either create score in view of nodes neighborhoods or edge information. Liben-Nowell and Kleinberg [1] altogether assessed a number of unsupervised methods and reasoned that the Adamic-Adar measure of node similarity performed best. They additionally found that basic common neighbor indicator worked surprisingly very much contrasted with Jaccard's coefficient, SimRank [2] and random walk-based hitting time. In reference [3], the author used a modified random walk approach, Customized PageRank [4] to ascertain rank parameter for every node.

The authors in [5] painstakingly clarified a portion of the properties of imbalance [6] in sparse systems with its relationship to network graph separation, and how to overcome this by directed learning. We finally accomplished some attractive outcomes by extracting, for example, in-degree, out-degree and some unsupervised measures, for example, number of basic neighbors, Adamic-Adar, shortest path as elements to prepare some troupe classifiers like Random Forests. Moreover, likeness scores were separated as components for supervised classification in the examinations of [7], which enriched the list of capabilities into a more far-reaching one with 39 different features for example, cosine similarity, Bayesian Sets, EdgeRank and so forth.


## 3 Methodology

The critical tasks in methodology incorporate information preprocessing assignments before applying any calculations for expectation and examination. Different strides in our present work incorporate readiness of node edge datasets from the interpersonal network graph, supporters network era, undesirable information expulsion for adjusting the dataset, feature extraction and so on. The whole design is appeared in (Fig. 1).


### 3.1 Overview

In our investigation, the preprocessed social network graph G is represented as an adjacency matrix framework a comprising of 4,867,136 clients joined by 20,000,000 edges as shown below. To foresee whether a specific testing edge is genuine or

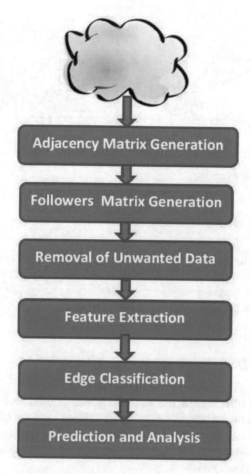

**Fig. 1** Architecture of link prediction

fake, we prepared the input dataset by consistently inspecting 10,000 positive edges and 10,000 negative sets in view of network A, which was principally because of the confinements of memory and processor. At that point, all of these edges are changed to a vector of components with a name of genuine or fake. Along these lines, our connection expectation and future client communication identification issue can be settled by utilizing unsupervised and supervised learning models with effective components. Adjacency matrix is created as per the following formula (Fig. 2).

$$A(i, j) = \begin{cases} 1, & \text{if there is an edge between nodes } v_i \text{ and } v_j \\ 0, & \text{otherwise} \end{cases}$$

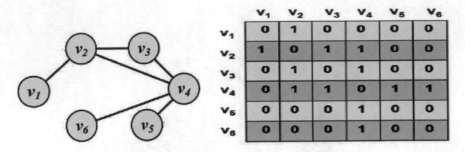

**Fig. 2** Adjacency matrix creation

## 3.2 Followers Matrix Computation

The preparation information given for tests is a tab-delimited contiguous network, where each column speaks to a client and its outbound neighbors (followees). Consequently, we pre-created a followers lattice B from the given unique graph, for obtaining followers measurements with less intricacy.

## 3.3 Celebrity Data Removal

In our preparation information, the normal number of followees for every client is 93, and henceforth nodes having more than 100 followees have been disposed of for the purpose of getting a more precise classifier. So positive edges testing and negative edges producing and testing has been done in prepared input data collection to make simplicity of computational investigation.

## 3.4 Positive Edges Sampling

Preparing 2,000,000 edges without a moment's delay would be incomprehensible because of the memory limitation. A regular option is perform preparing on simply part of the information, which can be accomplished by consistently examining. In our analysis, 10,000 positive edges are examined as a small amount of the final training data le.

## 3.5 Negative Edges Generation and Sampling

The first training data just furnishes us with positive edges, which implies negative connections should be created physically. In view of previously mentioned information set network A and B, negative edges are created in the following way: If client v is excluded in the followees set of client u, one might say that connection is negative. We haphazardly got client v 100,000 times to create 10,000 negative edges as another portion of the final training data. The most imperative impact is that when negative weights are available low-weight shortest paths have a tendency to have a larger number of edges than higher-weight paths. At the point, when negative weights are available, we look for makeshift routes that use the greatest number of edges with negative weights as we can discover.

## 3.6 Feature Set Extraction

In this report, we have normally investigated the utilization of three capabilities: proximity, ego-centric and aggregation [8]. We have considered principally three components in this examination. Proximity highlights are qualities that speak to some type of nearness among the combination of nodes [2]. Ego-driven elements are those elements focusing on the nearby system of u or v. An illustration would be the number of supporters of u. Aggregation is connected to produce effective components for interface expectation, which are a total of followees, whole of supporters, a total of companions and whole of neighbors.

## 3.7 Proximity Feature

Proximity highlights are qualities that speak to some type of closeness between the match of clients [9]. For example, it is exceptionally conceivable that a client could take after another with whom they share a common companion. Furthermore, vicinity highlights are typically shabby to be registered. Meanings utilized as a part of this segment are:

$\Gamma_{in}(x)$ represents the followers of x; $\Gamma_{out}(x)$ represents the followees of x.
Number of Common Followers: Represents the common set of the followers of x and y.
$NCFER(x,y) = |\Gamma_{in}(x) \cap \Gamma_{in}(y)|$
Number of Common Followees: Represents the common of the followees of x and y.
$NCFEE(x,y) = |\Gamma_{out}(x) \cap \Gamma_{out}(y)|$

Number of Common Clients: Represents the common set of clients between x and y, where bi-directional verges denotes friendship. It is expressed as: $NCC(x,y) = |\Gamma_{out}(x) \cap \Gamma_{in}(x)| \cap |\Gamma_{out}(y) \cap \Gamma_{in}(y)|$

Number of Common Neighbor: Measures the overlap of local networks of x and y by ignoring the directions of links. [2] This feature is included to evaluate the impact of direction on link creation.

$NCN(x,y) = |\Gamma_{out}(x) \cup \Gamma_{in}(x)| \cap |\Gamma_{out}(y) \cup \Gamma_{in}(y)|$

Matching Measurements: Measures the factual comparability of the followees, supporters, companions and neighbors among x and y. In our examination, three techniques are utilized, which are cosine similarity, Jaccard Coefficient and Adamic-Adar similitude [8], separately.

## 3.8   Ego-Centric Features

Ego-driven components are those elements focusing on the nearby system of u or v. A case would be the quantity of supporters of u. We trust that the commitment of this kind of elements can be exceptionally useful in connect forecasts. Specifically in our investigation, four ego-centric components are incorporated. These are number of adherents of u or v, the quantity of followees of u or v, the log proportion of u's adherents and v's devotees and the log proportion of u's followees and v's followees.

## 3.9   Aggregation Features

So as to make the features turn out to be more identified with both u and v, the least difficult accumulation function SUM is used to produce effective features for connection forecasting, which are aggregate of followees, total of supporters, entirety of companions, total of neighbors individually.

## 3.10   Edges Classification

Countless calculations can be decided for interface expectations. In this report, we played out a few trials on unsupervised learning strategies by utilizing previously mentioned likeness features [6] and regulated learning calculations including KNN, Random Forests and SVM. For the execution of the calculations, an outside machine learning bundle called WEKA has been utilized.

# 4 Unsupervised Learning

We began classification with unsupervised learning [10] since it is all the more straight-forward than supervised learning. To guarantee that unsupervised learning would create alluring outcomes, we have to apply a few features in a way that can reflect the general measurable example of the testing information. Specifically, the closeness score (Cosine, Jaccard and Adamic-Adar) of u and v would be processed and at that point scaled to [0, 1] as the confidence score of whether the connection between them is genuine or not. Mathematical formulation of Cosine, Jaccard and Adamic-Adar similarity measures is explained in the following subsections.

## 4.1 Cosine Similarity

Computation method works on the dot product of two vectors. In network link prediction, for every pair of nodes having common neighbor, the method performs a vector multiplication. It is calculated as

$$\frac{|\Gamma(x)||\Gamma(y)|}{\|\Gamma(x)\| * \|\Gamma(y)\|}$$

where $x$ and $y$ are two nodes.

## 4.2 Jaccard Similarity Coefficient

The Jaccard coefficient is a statistical measure utilized for contrasting similarity of test sets. In network link prediction, every one of the neighbors of a node is dealt with as a set and the prediction is finished by computing and positioning the closeness of the neighbor set of every node pair. It is calculated as

$$\frac{|\Gamma(x) \cap \Gamma(y)|}{|\Gamma(x) \cup \Gamma(y)|}$$

## 4.3 Adamic-Adar Index

Adamic-Adar is explained as the regular neighbor of a pair of nodes with few neighbors contributes more to the Adamic/Adar score (AA) esteem than this with substantial number of connections. In genuine social network, it can be translated as: If a typical colleague of two individuals has more companions, at that point, it is more improbable that he will acquaint the two individuals with each other than for the situation when he has just few friends. It is calculated as

$$\sum_{z \in \Gamma(x) \cap \Gamma(x)} \frac{1}{\log|\Gamma(z)|}$$

where $z$ is a common neighbor of node $x$ and node $y$.

## 5   Supervised Learning

In supervised learning, we conducted experiments using three calculations, which are KNN, Random Forests and non-direct SVM individually. KNN was picked up as our classifier competitor on the grounds that KNN is an exceptionally basic classifier that functions admirably on essential recognition issues. Moreover, KNN is hearty training data with noise and effective when the training data is huge in size. By applying bagging which is one of the ensemble methods, we expect that the execution would be significantly superior to the base classifier. With respect to non-linear SVM [11], even despite the fact that the training time is moderately long [12], it is equipped for catching complex connections between nodes with regard to interface forecast. Mathematical formulation of KNN, Random Forests and non-linear SVM are explained in the following subsections.

## 6   KNN

K-nearest neighbors, or KNN, is a group of straightforward arrangement based on similarity (Distance) calculation between instances. Nearest neighbor implements rote learning. It depends on a neighborhood average calculation. For few given data items for training and furthermore another unlabeled data item for testing purpose, our point is to discover the label of the class for the tested data item. The calculation has diverse behavior in light of $k$.

For $k = K$, we wish to discover the $k$ closest neighbor and check for larger part affiliation. Ordinarily $k$ is odd when the quantity of classes is 2. Lets say $k = 5$ and there are 3 samples of Class-1 and 2 samples of Class-2. Here, KNN says that new data item must belong to Class-1 because it has lion's share. We take a similar comparative contention for the problems having numerous classes. The closest neighbor strategy delivers a linear decision limit. It's somewhat more complex as it creates a piece-wise direct decision of the boundary with at times a group of minimal linear pieces.

## 6.1 Random Forest

Random Forest (RF) algorithm is an ensemble learning method for classification technique and can be used in social network data training, learning and classification tasks. This method form decision trees' rules of over fitting to their training set. For the replacement of non-categorical missing data value in the training set, RF computes the median of all values of variable $m$ in class $j$, and all the missing values are replaced by this computed value in class $j$. For the categorical $m$-th variable data, the replacement is done by using the most frequent non-missing value of the same class. In case of test dataset, missing values are replaced by the most frequent non-missing value derived from the training set. In unsupervised learning, the dataset comprises of a set of $x$-vectors of the same dimension with no class labels.

RF considers the original data as Class 1 and creates a synthetic same size class labeled as 2. The synthetic second class is created by sampling at random from the univariate distributions of the original data. For a single member class, it is converted as $x(1, n) \rightarrow x(2, n)$, where second coordinate is independently created from the $N$ values. Now all of the random forest computational aspects are applied over these two classes. The pseudocode for Random Forest is outlined below:

1. Select "$k$" features from given $n$ number of features randomly, provided $k < n$.
2. From $k$ number of features, a node "$d$" is selected using best split concept.
3. Again "$d$" is splitted using best split policy.
4. Steps 1–3 are repeated until "$m$" number of nodes are visited.
5. Steps 1–4 are repeated to create required number of trees to form the Random Forests.

For prediction using Random Forest, all the test features and randomly created decision trees are taken together to compute the predicted result. Votes are computed against each predicted target for final prediction in the entire random forest.

## 6.2 Non-linear SVM

Non-linear SVM performs binary classification using non-linear Support Vector Clustering with RBF kernel. The target to predict is a XOR of the inputs. Our goal is to discover a separating hyperplane that accurately discriminates the two classes.

A hyperplane can be written as

$$\vec{w} . \vec{x} - b = 0,$$

where $\vec{w}$ is normal vector and $\vec{x}$ is the set of points.

From the given dataset and mapping function, we can compute a decision function that separates the non-linear dataset, $f(\bar{x}) = \text{sign}(\bar{w}^* \cdot \text{Phi}(\bar{x}) - b^*)$

## 7  Experimental Results and Analysis

In both training and testing dataset, number of genuine edges and false edges were nearly the same, which implies that a pattern classifier would have a precision of half by anticipating all testing edges as 1 or 0.

Table 1 portrays the execution after effects of unsupervised learning on similarity features. It can be seen that all the comparability features we endeavored accomplished exactness over 80%, which shows that these comparability features have a decent capacity of separating genuine and fake connections.

Table 2 demonstrates the comparative execution of different supervised classifiers on training and testing dataset. The execution results in Table 1 were deduced by 10-overlay cross-approval on training data le and Table 2 was developed in light of the AUC (Area Under Curve) scores over testing dataset. As should be obvious from these two tables, non-linear SVM played out the best for both datasets with an exactness of 76.3 and 75.5%, separately. Also, all classifiers accomplished a superior execution on training data le than testing information, which uncovers that even in spite of the fact that cross-validation is connected to relieve the danger of overfitting, these classifiers still over-fitted training data to some degree. To think about the consequences of Tables 1 and 2, it can be shockingly discovered that unsupervised learning methods perform significantly superior to anything supervised learning models. Furthermore, it can be induced that wrong classifications are probably contributed by the parts sitting under some overlapping common regions for general features [13].

**Table 1**  Unsupervised classification results

| Similarity measure | Accuracy | Avg. accuracy |
|---|---|---|
| Cosine similarity | 81.00 | |
| Jaccard coefficient | 79.00 | 80.00 |
| Adamic-Adar similarity | 80.45 | |

**Table 2**  Supervised classification results

| Classifier's name | Accuracy level | Average accuracy |
|---|---|---|
| KNN | 73.00 | |
| Random Forest | 74.00 | 74.43 |
| SVM | 76.30 | |

# 8 Conclusion and Future Works

Connection forecast in huge network graph is still extremely difficult and hence draw the attention of many social network researchers. A considerable measure of research focus has been given around there. We discovered that in unsupervised space, cosine similarity performed best, took after by Adamic-Adar and afterward Jaccard's coefficient. In supervised space, we separated 14 features from the system structure and connected classifiers on them. To our astound, Random Forests did not accomplish the best execution but rather non-linear SVM outperformed it. The reasons were talked about in this report, and we feel that our testing and feature extraction should be enhanced for a superior result as far as supervised learning is concerned.

There were many promising methodologies which were not executed and tried because of equipment or time requirements. We might want to investigate more possibility in connect forecast research area in future. This examination proposes that clients with comparable topical interests will probably be companions, and subsequently semantic similarity measures among clients construct exclusively in light of their explanation metadata which ought to be prescient of social connections. This exploration work is proceeding to test more number of calculations both in supervised and in unsupervised [14] spaces and to find out more precise practical outcomes. We hereby acknowledge all the reference and helps taken from various sources including wikipedia, cited and all the uncited sources, if any, while preparing this manuscript.

# References

1. Liben-Nowell, D., Kleinberg, J.: The link prediction problem for social networks. In: Proceedings of the Twelfth Annual ACM International Conference on Information and Knowledge Management, New York, USA, pp. 556–559, Nov 2003
2. Jeh, G., Widom, J.: SimRank: a measure of structural-context similarity. In: Proceedings of 12 the ACM SIGKDD International Conference on Knowledge Discovery and Data Mining, Philadelphia, PA, USA, 20–23 Aug 2006
3. Narang, K., Leman, K., Kumaraguru, P.: Network ows and the link prediction problem. In: SNAKDD'13, NY, 2013
4. De Clercq, W., Vergult, A., Vanrumste, B., Van Paesschen, W., Van Hu el, S.: The PageRank Citation Ranking: Bringing Order to the Web. Technical report, Stanford University, Stanford (1999)
5. Lichtenwalter, R.N., Lussier, J.T., Chawla, N.V.: New perspectives and methods in link prediction. In: Proceedings of the 16th ACM SIGKDD International Conference on Knowledge Discovery and Data Mining, ACM, pp. 243–252, July 2010
6. Backstrom, L., Leskovec, J.: Supervised random walks: predicting and recommending links in social networks. In: Proceedings of the Fourth ACM International Conference on Web Search and Data Mining, ACM, pp. 635–644, Feb 2011
7. Cukierski, W., Hamner, B., Yang, B.: Graph-based features for supervised link prediction. In: The 2011 International Joint Conference on Neural Networks (IJCNN). IEEE, pp. 1237–1244, Jul 2011

8. Al Hasan, M., Chaoji, V., Salem, S., Zaki, M.: Link prediction using supervised learning. In: SDM6: Workshop on Link Analysis Counter-terrorism and Security, Oct 2006

9. Li, P., Liu, H., Yu, J.X., He, J., Du, X.: Fast single-pair simrank computation. In: SDM, pp. 571–582, Dec 2010

10. Rowe, M., Stankovic, M., Alani, H.: Who will follow whom? Exploiting semantics for link prediction in attention-information networks. In: The Semantic WebISWC, pp. 476–491. Springer, Heidelberg (2012)

11. Gupta, P., Goel, A., Lin, J., Sharma, A., Wang, D., Zadeh, R.: Wtf: the who to follow service at twitter. In: Proceedings of the 22nd international conference on World Wide Web, pp. 505–514. May 2013

12. Hotelling, H.: Survey on link prediction In Facebook and Twitter. Int. J Eng. Res. Appl. 2(5), 1631–1637 (2012)

13. Yang, Y., Lichtenwalter, R., Chawla, N.: Evaluating link prediction methods. Knowl. Inf Sys. 45, 1–32 2014

14. Correa, N.M., Adali, T., Yi-Ou, L., Vince, D.C.: Exploring Supervised Methods for Temporal LinkPrediction in Heterogeneous Social Networks. In: International World Wide Web Conference Committee (IW3C2), Florence, Italy, 18–22 May 2015

# Sentiment Prediction of Facebook Status Updates of Youngsters

Swarnangini Sinha, Kanak Saxena and Nisheeth Joshi

## 1 Introduction

Behavioral issues among youngsters are the common problem which needs to be tackled carefully. If a youngster is inclined toward a self-harm or psychosocial damage, then definitely his behavior can be considered as risky and needs timely help before it goes out of control. The significant factors that influence their behavior are the interpersonal relationship and contextual pressure.[1]

Our daily life is greatly influenced by the Internet and its technologies. The most popular among these technologies are social media sites which are new and the most preferred ways of communication. Youngsters use social media for sharing their views, photos, chatting and staying in touch with friends and family, etc. Facebook is most widely used social media with billions of users spread across the world. Because of static nature of the status updates, they form a useful database for the researchers. These databases can be used for various kinds of research work in domains like politics, product reviews, movie ratings, news, market surveys, strategy planning for new product launch, behavioral analysis of youngsters, psychological traits of people, criminal mentality, and many more.

Among the different social media available today, Facebook is the most sort after media especially used by a huge network of friends. It plays significant role

---

[1] https://www.ncbi.nlm.nih.gov/books/NBK53418/.

S. Sinha (✉) · N. Joshi
Department of Computer Science, Banasthali Vidyapith, Vanasthali, Rajasthan, India
e-mail: swarnangini@gmail.com

N. Joshi
e-mail: jnisheeth@banasthali.in

K. Saxena
Department of Computer Application, Samrat Ashok Technological Institute, Vidisha, India
e-mail: ks.pub.2011@gmail.com

© Springer Nature Singapore Pte Ltd. 2019
R. K. Shukla et al. (eds.), *Data, Engineering and Applications*,
https://doi.org/10.1007/978-981-13-6347-4_9

in understanding the human feelings very closely because of the unique features it provides like:

- Most of the users are youngsters ranging between 15 years and 30 years of age.
- It is a network of friends.
- The youngsters prefer to use this media to express themselves fearlessly without any parental pressure.
- Facebook is not showing any gender bias.
- The posts shared by the users are written in natural language. There is no restriction of the word limit.
- The posts remain there on the timeline for a quite longer period of time.
- The kind of posts, likes, pictures, and comments shared on this media are directly related to the emotional connect of the user.
- It provides a voluminous data for sentiment extraction and predicting the emotional well-being of the user.
- It also helps to find important psychological traits of the younger generation which may affect society at large.
- The profile information provided by the users apart from their status updates also helps to know more about the background of the user.
- The kind of friends a user has also provided an important information about him.

Affective data analysis is gaining momentum for the past few years as more and more youngsters are spending their time on social media. The abundance of data available on these websites provides useful database for research. Scientists are using this data for finding reasons behind growing psychological problems among the youth. Generally, all these researches are available in English language. For the past few years, contents written in Hindi language are also increasing at a faster rate on the social media. But a very few researches related to opinion mining of product reviews, news, movies, etc. are available in Hindi language [1, 2, 3]. We have selected Facebook status updates of young people written in both Hindi and English languages as the domain of our study.

The proposed research is intended to find the emotional polarity of youngsters through these posts over a specific period of time. If the posts are found to be consistent, then it is considered as the normal behavior otherwise after detecting the emotion polarity as negative, we assume that user is going through a psychological problem. The remainder of this paper is organized as follows. Section 2 presents the literature review on affective data analysis and role of social media. Proposed methodology is described in detail in Sect. 3. Section 4 discusses the analysis of results. Finally, Sect. 5 concludes the work.

## 2 Literature Review

Affective data analysis is the process of detecting emotional polarity from the text, if it is positive, negative, or neutral. Emotions play a vital role in our day-to-day life and act as a driving force to lead our life in a particular direction. They help us to communicate, learn new things, and influence our decision-making ability [4]. With the proliferation of various social media, a tremendous amount of data is generated and resulted into gigantic datasets. It is a rich reservoir of a multimodal datasets in the form of text, audio, and video. The techniques which are mainly used to perform affective data analysis are bag-of-words, lexicon analysis, statistical approaches, and rule-based techniques [5, 6]. In the recent past, the scientific community has expanded their area of study by incorporating the emotions extracted from various kinds of text data gathered from the news, blogs, Twitter messages, Facebook posts, and customer reviews [7, 8, 9, 10].

S. Poria et al. proposed an intelligent framework for extracting information from multimodal sources [5]. With the help of affective content recognition system of natural language processing, it is possible to analyze the textual data. More and more research works are inclined toward this field because of the involvement of big companies for the online review of their products. The work was mainly centered around the identification of the emotional polarity in the form of positive, negative, and neutral sentiments associated with words [7, 11] and documents [12]. The automatic identification of the emotions such as suicidal tendencies, fear, anxiety, happiness, anger, sadness, and love hidden in these text data is studied by the research community for the past few years.

T. Goldfinch et al. proposed a semi-hierarchical knowledge framework for the students of engineering mechanics. The authors have suggested that the educators should find different ways of assessing the students of the first year so that they are encouraged to study concepts [13]. According to a research report of Department for Work and Pensions, extensive surveys have already been started to investigate social media data to find the probabilities of human behavior prediction. In contrast to traditional methods, these new methods offer the high level of correctness in predicting the results and are disseminated with very high speed [14].

R. Lin and Sonja Utz investigated an emotional state of people belonging to two different geographical areas through Facebook posts and studied relationship closeness existing between a user and Facebook post in predicting happiness and envy. When browsing Facebook, they found posts which are more inclined toward positive sentiments than negative feelings [15]. On examining the association between the properties of the Facebook profile and the personality distinctiveness of users, Y. Bachrach et al. established a close relationship between them [16]. The use of social media helps to spread emotional state from person to person. A. D. I. Kramera et al. stated that the feelings shared by others on Facebook manipulate our own sentiments. They proved this phenomenon with the help of an experiment carried out using large real-world social network [17].

N. Mittal et al. proposed a use of Hindi SentiWordNet for classifying Hindi movie reviews [18]. A. Bakliwal et al. suggested a graph-based WordNet expansion method to generate a full subjective lexicon for product review [1]. A. Joshi et al. studied fall-back strategy for classifying movie reviews [19]. N. Mittal et al. suggested proper handling of negation and discourse relation improves performance of sentiment analysis of Hindi reviews [20].

The task of identifying emotional polarity in Hindi language possesses some challenges because of the following reasons [21, 2, 20, 22].

- Scarcity of resources like Hindi language tagger, parser, and annotated corpus.
- Non-availability of standard datasets in Hindi language.
- Hindi is a free word order language unlike English.
- Same word may occur in multiple contexts.

Our study is a novel approach of finding emotional polarity of adolescents from their Facebook posts written in both Hindi and English languages. To the best of our knowledge, no previous study has been found in this domain for bilingual posts.

## 3   Proposed Methodology

We propose a framework which includes data collection, preprocessing, sentiment identification, feature extraction, sentiment classification, and sentiment polarity prediction. In order to perform our experiment, we used Weka machine learning toolkit, version 3.8. The dataset is divided into training dataset with 60% data and test dataset with 40% data. The training dataset is trained using different classifiers, and results are tested on test dataset for predicting the outputs. The architecture of the proposed system is shown in Fig. 1.

### 3.1   Data Collection

We have collected 5439 Facebook status updates over a period of three months from July 2015 to September 2015. The posts collected are bilingual written in both Hindi and English languages. Data are collected manually from the timeline and stored in a database which includes date, time, and the actual contents of the posts.

### 3.2   Data Preprocessing

The data collected are noisy and unstructured which cannot produce desired results. Hence, before applying any classification model on this data, it is subjected to preprocessing which includes the following steps.

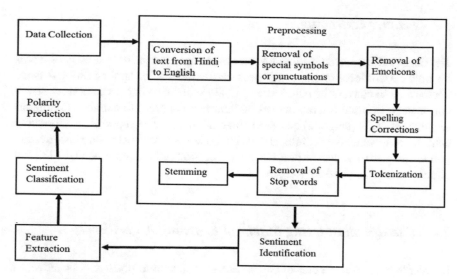

**Fig. 1** Proposed architecture

- Transliteration of status updates written in Hindi language to English language [2].

  बिना #तड़के_ की दाल_ और बिना #attitude_वाला_माल हमे बिलकुल पसंद नहीं है.

  bina #tadake_kee daal_aur bina #attitude_vaala_maal hame bilakul pasand nahin hai.

- Removal of unnecessary characters and punctuations that do not have any effect on the sentiment classification.

  #status__chahiye to maang liya kro…

  Status chahiye to maang liya kro

- Unnecessary spaces between words and characters are removed.
- Status updates are converted to lowercase.

  #kaMzOr__dIL__ki_lAdKiA MeRi timEliNe Na DekhE :-(

  #kamzor__dil__ki_ladkia meri timeline na dekhe :-(

- Removal of emoticons.
- The text data collected was written in slag language with lots of misspellings. Hence, spelling corrections are also made.
- Tokenization, i.e., conversion of data into a matrix of word vectors where each word represents a separate feature.
- Removal of stop words followed by stemming. A custom list of stop words is prepared for including stop words from both the languages like 'and,' 'or,' 'is,' 'ka,' 'ki,' 'ko,' etc.

## 3.3   Feature Extraction

The dataset after converting into a vector of strings contained thousands of words. So, only the words expressing sentiments were extracted. We applied StringToWord-Vector filter to convert the entire dataset (5439 sentences) into a vector of words. In our context, we used $N$-gram model (unigram and bigram) to extract a continuous sequence of 1672 (unigram) and 4281 (bigram) words. The supervised information gain attribute evaluator in combination with Ranker was used as the dimension reduction techniques to select the best feature words, 462 (with unigrams) and 373 (with bigrams) according to their weights.

## 3.4   Classification Using Machine Learning Algorithms

In order to predict sentiments of Facebook status updates as positive or negative, we used six standard text classification algorithms, namely Naive Bayes (NB), multinomial Naïve Bayes (MNB), Rule-PART, Tree-J48, $k$-nearest neighbor lazy-IBk and Functions(SMO), i.e., support vector machine, and evaluated their performance with different combinations of unigrams and bigrams.

**Naive base**: The algorithm is based on Bayes' theorem where each attribute is treated as independent term. It is a simple probabilistic algorithm. In our case, likelihood of attribute $(x)$ to be a member of the class $(c)$ in a document $(d)$ is given as

$$P(x_i|c) = \text{frequency of attribute in class/total attributes of a class}$$

According to Bayes' theorem, probability of a document belonging to a class $(c_i)$ is given as

$$P(c_i|d) = P(d|c_i) \times P(c_i)/P(d)$$

Hence, the maximum posterior probability of the class can be given as

$$P(c_i|d) = \left( \prod P(x_i|c_i) \right) \times P(c_i)/P(d)$$

where $x_i$ is individual word of the document.

**Multinomial Naïve Bayes**: It is a variation of Naïve Bayes model where term frequency is used to compute the maximum-likelihood estimate based on the training data to estimate the class-conditional probabilities.

$$P(x_i|c_j) = \sum tf(x_i, d \in c_j) + \alpha / \sum N_{d \in cj} +_\alpha \cdot V$$

where

- $x_i$: a word from the feature vector $x$ of a particular document.
- $\sum tf\ (x_i, d \in c_j)$: the sum of raw term frequencies of word $x_i$ from all documents in the training dataset that belong to class $c_j$.
- $\sum N_{d \in c_j}$: the sum of all term frequencies in the training dataset for class $c_j$.
- $\alpha$: an additive smoothing parameter ($\alpha = 1$ for Laplace smoothing).
- $V$: the size of the vocabulary (number of different words in the training dataset).

The class-conditional probability of encountering the word $(x)$ can be calculated as the product from the likelihoods of the individual words (under naive assumption of conditional independence).

$$P(x|c_j) = P(x_1|c_j) \cdot P(x_2|c_j) \dots P(x_n|c_j) = \prod_{i=1}^{m} P(x_i|c_j)$$

**Rule**: Due to the small sample size, we have used rule-based system to categorize the textual document. PART rules for classification make use of separate and conquer technique and generate a partial C4.5 decision tree in each iteration to select the best leaf into a rule.

**Decision tree**: The algorithm generates pruned C4.5 decision tree. The main attributes, in the vector form of the text document, are selected depending on their weights, to form the tree. The algorithm relies on greedy search and builds a decision tree with top-down and recursive approach.

**K-nearest neighbor**: The appropriate value of $K$ based on cross-validation was selected. The model used weighted numeric values of the attributes for evaluation.

The presence or absence of feature vectors with 1-nearest neighbor was checked.

**Support vector machine**: This algorithm normalizes all attributes by default. It implements sequential minimal optimization algorithm for training a support vector classifier. It constructs hyperplane in a multidimensional space which divides the input data into different class labels. It applies an iterative training algorithm to minimize an error function and constructs an optimal hyperplane.

# 4 Results and Discussion

The dataset showed important characteristic features of the youngster's status updates which are mentioned as follows.

1. The status updates of young people are centered around themselves, friends, love life, activities happening in their daily life, and uploading of images about themselves. Younger people susceptible to negative mental status are self-obsessed and generally post data which reflect negativity.

2. They usually update status anytime during the day, but the frequency of posts increases during weekends, holidays, or at night. Youngsters with low self-esteem used to upload data more frequently during night as compared to others.
3. The data uploaded by young people with positive mental health keep the length of their posts precise and to the point. On the other hand, youngsters who are going through some mental trauma usually post emotional data at greater length with contents centered around love life, crime, or suicidal tendencies.

We trained the dataset for different classifiers using tenfold cross-validation. The dataset is partitioned into training set and test set. Out of 10 equal size subsamples, nine subsamples were used for training the model, and one was used to test the model. The results were averaged to produce final estimation.

The results obtained after applying classification algorithms with unigram and bigram models are given in Table 1.

Unigram: The accuracies of these classifiers have given very favorable results when used with unigrams. SMO has shown higher accuracy than the rest of the classifiers.

Bigram: The results obtained clearly indicate that the accuracies of multinomial Naïve Bayes and IBk have improved with the use of bigrams. On the other hand, SMO Naïve Bayes, PART, and J48 have shown decline in performance when used with bigrams. Though bigrams are equally important as unigrams in text analysis, in our context, it reduced the performance of the classifiers except for multinomial Naïve Bayes and IBk.

Figure 2 represents the comparative analysis of performance of these classifiers when used with unigram and bigram models.

Precision, recall, and $F$-measure of different classifiers also generate significant results as given in Table 2.

**Table 1** Comparative analysis of different classifiers

| Classifier | SMO | MNB | IBk | PART | J48 | NB |
|---|---|---|---|---|---|---|
| Unigram accuracy in % | **96.01** | **95.05** | **93.91** | 95.09 | 94.99 | 89.44 |
| Bigram accuracy in % | 94.96 | **95.40** | **93.96** | 86.67 | 87.51 | 81.85 |

**Table 2** Comparative analysis for accuracy features of different classifiers

| Classifier | | SMO | MNB | IBk | J48 | PART | NB |
|---|---|---|---|---|---|---|---|
| Unigram | Precision | 0.96 | 0.95 | 0.94 | 0.95 | 0.95 | 0.90 |
| | Recall | 0.96 | 0.95 | 0.94 | 0.95 | 0.95 | 0.89 |
| | $F$-measure | **0.96** | **0.95** | **0.94** | 0.95 | 0.95 | 0.89 |
| Bigram | Precision | 0.95 | 0.96 | 0.94 | 0.90 | 0.87 | 0.82 |
| | Recall | 0.95 | 0.95 | 0.94 | 0.88 | 0.87 | 0.82 |
| | $F$-measure | **0.95** | **0.95** | **0.94** | 0.88 | 0.86 | 0.81 |

The results in Table 2 show that MNB performs the best. SMO, MNB, and IBk perform well with high precision, recall, and $F$-measure rates over 0.9. J48, NB, and PART also gave competent results of precision, recall, and $F$-measure rates over 0.9, but their performance declined with the use of bigrams. NB performed worst with bigrams.

Performance of the best three classifiers is further represented using confusion matrix generated for neutral, positive, and negative classes as given in Tables 3, 4, and 5. The matrix shows how the classifier predictions are distributed, and how well the classifier learned to distinguish among various classes.

SMO, MNB, and IBk have shown 99% correct predictions of positive posts when used with unigram feature. While more than 85% negative posts were correctly classified by SMO and MNB. On the other hand, IBk could give 73% correct results. All classifiers have shown more even distribution of positive posts than negative or

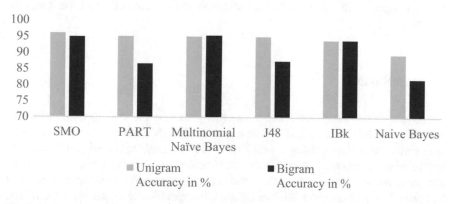

**Fig. 2** Performance analysis of classifiers

**Table 3** Confusion matrix of SMO classifier

| SMO classifier | Unigram | | | Bigrams | | |
|---|---|---|---|---|---|---|
| | Neutral | Positive | Negative | Neutral | Positive | Negative |
| Neutral | 752 | 112 | 4 | 729 | 139 | 0 |
| Positive | **16** | **3932** | **12** | 1 | **3956** | **3** |
| Negative | 12 | 61 | 538 | 1 | 130 | 480 |

**Table 4** Confusion matrix of multinomial Naïve Bayes classifier

| MNB | Unigram | | | Bigrams | | |
|---|---|---|---|---|---|---|
| | Neutral | Positive | Negative | Neutral | Positive | Negative |
| Neutral | 743 | 115 | 10 | 718 | 145 | 5 |
| Positive | **43** | **3901** | **1** | **16** | **3938** | **6** |
| Negative | **14** | **71** | **526** | **9** | **69** | **533** |

**Table 5** Confusion matrix of IBk classifier

| IBk classifier | Unigram | | | Bigrams | | |
|---|---|---|---|---|---|---|
| | Neutral | Positive | Negative | Neutral | Positive | Negative |
| Neutral | 722 | 143 | 3 | 718 | 150 | 0 |
| Positive | **17** | **3940** | **3** | 2 | **3955** | **3** |
| Negative | 50 | 115 | 446 | 0 | 173 | 438 |

neutral posts. The reason behind this could have been the presence of more positive posts than the others.

The performance of SVM declined to 78% and IBk decreased to 71% when used with bigrams, while MNB outperformed with an increase of 87.23%. Hence, it can be concluded that performance of SVM and IBk reduces with an increase in the sparsity degree of feature vectors. On the contrary, it does not affect the performance of MNB.

## 5 Conclusion

We have collected Facebook status updates of young people to study their emotional polarity. The dataset is trained to automatically classify emotions as positive, negative, and neutral classes using six different classifiers. We observed that incorrectly classified sentences were the result of complex language structure and manual interpretation of class labels. In the future, the accuracy of classification can be improved by taking into account more NLP techniques and specifically improving class labels.

## References

1. Bakliwal, A., Arora, P., et al.: Hindi subjective lexicon: a lexical resource for hindi polarity classification. In: 8th International Conference on Language Resources and Evaluation (LREC) (2012)
2. Ghosh, A., Dutta, I.: Real-time sentiment analysis of hindi tweets. In: International Conference of the Linguistic Society of Nepal (2014)
3. Jha, V., Manjunath, N., Shenoy, P.D., Venugopal, K.R.: Sentiment analysis in a resource scarce language: Hindi. Int. J. Sci. Eng. Res. 7(9) (2016)
4. Howard, N., Cambria, E.: Intention awareness: improving upon situation awareness in human-centric environments. Human-Cent. Comput. Inf. Sci. 3(9) (2013)
5. Poria, S., Gelbukh, A., Hussain, A., Das, D., Bandyopadhyay, S.: Enhanced SenticNet with affective labels for concept-based opinion mining. IEEE Intell. Syst. 28(2), 31–38 (2013)
6. Xia, R., Zong, C.Q., Hu, X.L., Cambria, E.: Feature ensemble plus sample selection: a comprehensive approach to domain adaptation for sentiment classification. IEEE Intell. Syst. 28(3), 10–18 (2013)

7. Balahur, A., Hermida, J. M., Montoyo, A.: Building and exploiting emotinet, a knowledge base for emotion detection based on the appraisal theory model. IEEE Trans. Affect. Comput. **3**(1) (2012)
8. Cambria, E., Olsher, D., and Rajagopal, D.: SenticNet 3: a common and common-sense knowledge base for cognition-driven sentiment analysis. In: AAAI, pp. 1515–1521 (2014)
9. Pak, A., Paroubek, P.: Twitter as a corpus for sentiment analysis and opinion Mining. In Proceedings of LREC (2010)
10. Sidorov, G., Miranda-Jiménez, S., Viveros-Jiménez, F., Gelbukh, A., Castro-Sánchez, N., Velásquez, F., et al.: Empirical study of machine learning based approach for opinion mining in tweets. Lecture Notes in Artificial Intelligence, 7629, pp. 1–14 (2013)
11. Wawer, A.: Extracting emotive patterns for languages with rich morphology. Int. J. Comput. Linguist. Appl. **3**(1), 11–24 (2012)
12. Maas, A., Daly, R., Pham, P., Huang, D., Ng, A., Potts, C.: Learning word vectors for sentiment analysis. In: Proceedings of the association for computational linguistics, ACL, Portland (2011)
13. Goldfinch, T., Carew, A. L., McCarthy, T.J.: A knowledge framework for analysis of engineering mechanics exams. In: Research in Engineering Education Symposium VIC, Australia: Melbourne School of Engineering, The University of Melbourne, pp. 1–6 (2009)
14. The Use of Social Media for Research and Analysis: A Feasibility Study. Report of Department for Work and Pensions (2014)
15. Lin, R., Utz, S.: The emotional responses of browsing facebook: happiness, envy, and the role of tie strength. Comput. Human Behav. **52**, 29–38 (2015). Elsevier
16. Bachrach, Y., Kosinski, M., Graepel, T., Kohli, P., Stillwell, D.: Personality and patterns of facebook usage. In: Proceedings of Annual ACM Web Science Conference, ACM (2012)
17. Kramera, A.D.I., Guilloryb, J.E., Hancock, J.T.: Experimental evidence of massive-scale emotional contagion through social networks. PNAS, vol. 111 (2014)
18. Mittal, N., Agarwal, B., Chouhan, G., Pareek, P., Bania, N.: Sentiment analysis of Hindi review based on negation and discourse relation. In: 11th Workshop on Asian Language Resources (ALR), In Conjunction with IJCNLP (in press)
19. Joshi, A.R., Balamurali, P.: A fall-back strategy for sentiment analysis in Hindi: a case study. In: International Conference on Natural Language Processing, ICON (2010)
20. Mittal, N., Agarwal Basant, et al.: Discourse based sentiment analysis for Hindi reviews. In: International Conference on Pattern Recognition and Machine Intelligence pp. 720–725 (2013)
21. Arora, P., Bakliwal, A., Varma, V.: Hindi subjective Lexicon generation using WordNet graph traversal. Int. J. Comput. Linguist. Appl. **3**(1), 25–39 (2012)
22. Sharma, R., Nigam, S., et al.: Polarity detection of movie reviews in Hindi language. Int. J. Comput. Sci. Appl. **4**(4), 49–57(2014)

# Part II
# On Machine Learning

# Logistic Regression for the Diagnosis of Cervical Cancer

Siddharth Singh, Shweta Panday, Manjusha Panday
and Siddharth S. Rautaray

## 1 Introduction

Generally, cervical cancer is caused by genital human papillomavirus (HPV) infection. This virus mostly spreads through sexual contact. Usually, female bodies are able to fight this. But the persistent HPV viruses may cause changes in cervical cells which are a major reason for cervical cancer. Out of 12,000 diagnosed cases of cervical cancer, 4000 die each year. This is the scenario of a developed country; from here, we can imagine the condition of developing nations.

Cervical cancer is a type of cancer which happens due to abnormal growth of cell in the cervix part of human body. This type of cancer can be treated easily if found in early stages. Anyone can get a cervical cancer if they come in contact with HPV virus. But all HPV viruses are not dangerous or show early symptoms. Some cause genital warts also (Fig. 1).

Cervical cancer mostly occurs during midlife of a person (20 and above). 15% of the patients with cervical cancer are of age from 50 to 65, as many people do not realize that as they age the chances of getting cervical cancer do not decrease. As in any other diseases, the existence of multi-screening and diagnosis methods creates a complex ecosystem from a computer-aided diagnosis (CAD) system point of view. For instance, in the detection of pre-cancerous cervical lesions, screening strategics include cytology, colposcopy (covering its several modalities [1]), and the gold-standard biopsy. This makes people dependent on doctor's knowledge and experience for the diagnosis of cancer. If we see this from a technical view, this methodology will have a multi-modal and multi-expert setting, which is not only time-consuming but also a complex process.

S. Singh (✉) · S. Panday · M. Panday · S. S. Rautaray
School of Computer Engineering, KIIT University, Bhubaneswar, India
e-mail: siddharth96@zoho.com

S. Panday
e-mail: shweta.sunil16@gmail.com

© Springer Nature Singapore Pte Ltd. 2019
R. K. Shukla et al. (eds.), *Data, Engineering and Applications*,
https://doi.org/10.1007/978-981-13-6347-4_10

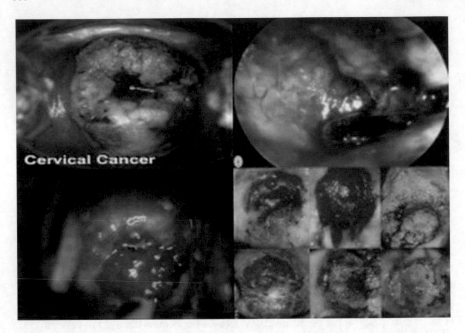

**Fig. 1** Images for cervical cancer. *Source* Google

Hinselmann's test is a testing given by Hans Hinselmann in the year 1924. He developed a colposcope prototype which will help in recognition of cervical cancer in early stages which was not possible prior to his time. Now, a colposcopy test is done after a pap test to check the existence of tumor in cervix. Before Hinselmann, the test for cervical cancer consists of palpation and doctor's knowledge. Hinselmann gave us approach which magnifies the view of cervix, and in return we get a more accurate result. This is done using a colposcope which initially made of lenses, binoculars, a stand, light source, and a couple of screws.

Now, the good part is it is curable. There are various methods to cure cervical cancer. Although this paper does not deal with curation of cervical cancer, there are a few of methods from which we can cure cervical cancer. For various stages of cancer, there are different treatments. Be it chemotherapy or radiation therapy. In some cases, surgery is also required. But there is a slight trouble if cancer is not found in early stages then it may require hysterectomy. Now, after this the patient or the person suffering from cervical cancer cannot have children. Hence, finding it in early stage is quite important. This paper is solely focused on early diagnosis of cervical cancer with the help of technology to make the process easy.

During this process, a woman is led on to the examination table with bare lower half. Then with the help of a smear, the walls of vagina will be spread, which further will make cervix visible. Then with the help of a brush, some cells are removed from the cervix. These cells are kept safely for further diagnosis under microscope, which in the end will give the result about existence of abnormalities in your cervix. During

  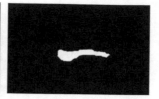

**Fig. 2** Images of the diagnosis (original, artifacts, os)

the process, a person may feel stomach cram or blood, but it is nothing to worry about.

Transfer learning (TL) aims to extract knowledge from at least one source task and use it when learning a predictive model for a new target task [2]. The initial idea behind this is that learning a new thing from the old one would be easier. For example, if we have studied the difference between the rotten potato and a fresh potato, then we can use the same reference to deduct between a rotten tomato and a fresh tomato (source: Quora). In this work, we focus on inductive TL, where both domains are represented by the same feature space and where the source task is different from the target tasks and yet they are related [2]. The main trend in inductive transfer consists of transferring data, namely strategically including data from the source task in the target dataset [3]. Another approach consists of finding a shared source-target low-dimensional feature representation that is suitable for learning the target task [4]. We combine these two and use it to get the required data (Fig. 2).

Logistic regression technique is a widely accepted for the statistical modeling; here the probability of the outcome is dependent on the series of predictor variables:

$$\log\left(\frac{p(x)}{1 - p(x)}\right) = \beta_0 + x_1 \cdot \beta_1 + x_2 \cdot \beta_2 + \cdots + x_n \cdot \beta_n$$

where $p$ is the probability of person suffering from cancer, $\beta_0$ is the probability of person suffering from cancer when all other variables from $x_1 \ldots x_n$ are zero that is intercept. $\beta_1 \ldots \beta_n$, are the $\beta$ coefficients for each variable $x_1, \ldots, x_n$.

On solving for $P(x)$, we will get

$$P(x) = \frac{e^{\beta_0 + x_1 \cdot \beta_1 + x_2 \cdot \beta_2 + \cdots + x_n \cdot \beta_n}}{1 + e^{\beta_0 + x_1 \cdot \beta_1 + x_2 \cdot \beta_2 + \cdots + x_n \cdot \beta_n}} = \frac{1}{1 + e^{-(\beta_0 + x_1 \cdot \beta_1 + x_2 \cdot \beta_2 + \cdots + x_n \cdot \beta_n)}}$$

Once that is done, we optimize the likelihood of the data with respect to the parameter $\beta$. The definition of likelihood is that is the measure of probability of data in provided setting. Here the data is fixed

$$L(\beta_0, \beta) = \prod_{i=1}^{n} p(x_i)^{y_i} (1 - p(x_i))^{(1-y_i)}$$

Later on we simplify this equation using log

$$l(\beta_0, \beta) = \sum_{i=1}^{n} (y_1 \log P(x_i) + (1 - y_i)\log(1 - P(x_i)))$$

$$= \sum_{i=1}^{n} \log(1 - p(x_i)) + \sum_{i-1}^{n} y_i \log \frac{P(x_i)}{1 - P(x_i)}$$

$$= \sum_{i=1}^{n} \log(1 - P(x_i)) + \sum_{i=1}^{n} y_i (\beta_0 + x_1 \beta)$$

$$= \sum_{i=1}^{n} -\log(1 + e^{\beta_0 + x_i \cdot \beta}) + \sum_{i=1}^{n} y_i (\beta_0 + x_i \cdot \beta)$$

After this, we optimize the likelihood by taking the derivatives with respect to $\beta$ and later on equating it with zero.

$$\frac{\partial l}{\partial \beta_j} = -\sum_{i=1}^{n} \frac{1}{1 + e^{(\beta_0 + x \cdot \beta)}} e^{(\beta_0 + x \cdot \beta)x_{ij}} + \sum_{i=1}^{n} y_i x_{ij}$$

$$= \sum_{i=1}^{n} (y_i - p(x_i; \beta_0, \beta)) x_{ij} = 0$$

Now, the basic question arises that—why this method and not any other method. This can be answered very easily by stating two major differences—one, conditional distribution is a form of Bernoulli distribution, other—the probability of predicted values varies between (0, 1) which is possible in case of logistic distribution. In simple words, this method can give us more accurate result in this case than any other at any particular time. This is one of the reasons why we chose this method over other methods.

In our work, we are using the transfer learning tools and techniques to give a more accurate result for the diagnosis of cervical cancer. As it is said, if we diagnose this cancer in the earlier stage, then we can cure it very easily. This can also help in effectively reducing the mortality caused by cervical cancer in future.

## 2 Methods

### 2.1 Previously Proposed Model

In the previously proposed model, they were converting the colposcopy images into various datasets which were sent to a panel of doctors for the diagnosis. This method was time-consuming as they have to consider a multi-modal as well as a multi-expert approach for the effective output.

## 2.2 Our Model

The approach we have taken is logistic regression; this is an upgraded version of the previously proposed paper. Here instead of using multi-expert setting, we are using a logistic approach to the number of dependent variables which can simplify the diagnosis approach. On comparison, our method gives more than 99% success rate in predicting whether a person is suffering from cervical cancer or not.

## 2.3 Experiments

The variables used here are 63 (62 predictive attributes, 1 target attribute). Out of these 63 variables, we are finally reducing it to 18 significant variables. This will help us in giving us an effective reading from our dataset. Using the correlation between the variables, we have applied step-wise regression for the reduction of variables in the diagnosis process. We have used one categorical variable in this which will give us the final output. The 18 variables are shown in the image (Fig. 3).

The approach we have used for the reduction of variables is logistic regression. Here at first, a generalized logistic model was plotted then using step-wise regression we kept on removing variables. At the end, we will have a list of variables which are significant and play a vital role in the diagnosis of cervical cancer (Fig. 4).

The dataset was acquired and annotated by professional physicians at 'Hospital Universitario de Caracas' (Source: UCI machine). Data was split into 70–30 with the help of a stratified training test partition. Then we used the 70% of the data for training our model and the rest 30% of data for the verification. We have gotten a 100% accuracy rate in our training dataset (Fig. 5).

Here we are trying to eliminate the number of dependent variables and make the process of diagnosis simpler and effective. We are also reducing the multi-expert factor which was previously in use. Therefore, I think our model, if comes into work, will be a major boost in reducing the deaths due to cervical cancer.

Coefficients:

|  | Estimate | Std. Error | z value | Pr(>\|z\|) |
|---|---|---|---|---|
| (Intercept) | -24722.64 | 654468.10 | -0.038 | 0.970 |
| walls_area | -4230.63 | 142741.46 | -0.030 | 0.976 |
| artifacts_area | 10109.60 | 437662.23 | 0.023 | 0.982 |
| walls_artifacts_area | -2720.34 | 77170.74 | -0.035 | 0.972 |
| speculum_artifacts_area | -3456.91 | 185382.88 | -0.019 | 0.985 |
| walls_specularities_area | -130054.37 | 3963581.87 | -0.033 | 0.974 |
| rgb_cervix_g_mean | -3408.17 | 96151.16 | -0.035 | 0.972 |
| rgb_total_r_std | -146.22 | 4515.68 | -0.032 | 0.974 |
| rgb_total_g_mean | 5089.03 | 136288.07 | 0.037 | 0.970 |
| hsv_cervix_h_std | 17360.93 | 461015.91 | 0.038 | 0.970 |
| hsv_cervix_s_mean | -97.59 | 2599.67 | -0.038 | 0.970 |
| hsv_cervix_v_mean | 3410.50 | 96294.03 | 0.035 | 0.972 |
| hsv_total_h_std | -11323.26 | 303098.01 | -0.037 | 0.970 |
| hsv_total_s_mean | 103.69 | 2764.06 | 0.038 | 0.970 |
| hsv_total_v_mean | -5049.34 | 135241.64 | -0.037 | 0.970 |
| fit_cervix_hull_rate | -12570.74 | 431064.97 | -0.029 | 0.977 |
| fit_cervix_hull_total | -32680.37 | 940401.78 | -0.035 | 0.972 |
| fit_cervix_bbox_rate | 25051.53 | 717084.86 | 0.035 | 0.972 |
| fit_cervix_bbox_total | 26142.22 | 732063.60 | 0.036 | 0.972 |

(Dispersion parameter for binomial family taken to be 1)

    Null deviance: 1.2232e+02  on 97  degrees of freedom
Residual deviance: 4.5426e-06  on 79  degrees of freedom
AIC: 38

Number of Fisher Scoring iterations: 25

**Fig. 3** Significant variables

```
Call:
glm(formula = consensus ~ walls_area + artifacts_area + walls_artifacts_area +
    speculum_artifacts_area + walls_specularities_area + rgb_cervix_g_mean +
    rgb_total_r_std + rgb_total_g_mean + hsv_cervix_h_std + hsv_cervix_s_mean +
    hsv_cervix_v_mean + hsv_total_h_std + hsv_total_s_mean +
    hsv_total_v_mean + fit_cervix_hull_rate + fit_cervix_hull_total +
    fit_cervix_bbox_rate + fit_cervix_bbox_total, family = binomial,
    data = green_n)

Deviance Residuals:
      Min         1Q      Median         3Q         Max
-8.352e-04  -2.000e-08   2.000e-08   2.000e-08   7.451e-04
```

**Fig. 4** Logistic model

**Fig. 5** Resulting confusion
matrix

```
Confusion Matrix and Statistics

                    Reference
        Prediction  0  1
                 0  4  0
                 1  0 26


                    Accuracy : 1
                      95% CI : (0.8843, 1)
        No Information Rate : 0.8667
        P-Value [Acc > NIR] : 0.01366


                       Kappa : 1
     Mcnemar's Test P-Value : NA


               Sensitivity : 1.0000
               Specificity : 1.0000
            Pos Pred Value : 1.0000
            Neg Pred Value : 1.0000
                Prevalence : 0.1333
            Detection Rate : 0.1333
      Detection Prevalence : 0.1333
         Balanced Accuracy : 1.0000

           'Positive' Class : 0
```

# 3   Risk

## 3.1   Privacy Factor

Here we are predicting the presence of cervical cancer through various treatments
of dataset. These datasets are made by collecting data from various patients; now
here lays a valid chance of patients withholding the information due to privacy or
some other active factor. This also gives rise to another query which is whether they
are giving correct answers or not. As in the case of the collected data is not correct,
then the prediction of the diagnosis of the person regarding, whether the person is
suffering from cancer will also be ambiguous.

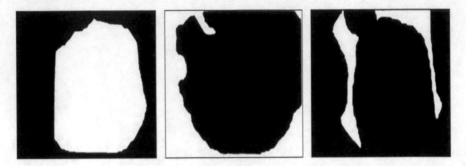

**Fig. 6** Images of cervix, speculum, and vaginal walls

## 3.2 Quality of Images

Choosing a good frame for the screening is one of the major requirements for the effective diagnosis of cervical cancer. As we are converting the images into datasets for better results. The primary source should be good and clear for an effective outcome. If the images are not clear, then a fatal error in the diagnosis process will occur, which can harm the person instead of helping them. So the most important part of diagnosis will be getting accurate and clear images (Fig. 6).

## 3.3 Mechanical Errors

We also need to consider the error due to machines. As most of the machines do not give 100% accuracy, there is a very small amount of error percent present. During the calculation for the diagnosis, we always need to consider this error.

## 3.4 Presence of Outliers

There are some cases of miracle and some are of curses. Sometimes, we hear a person has recovered miraculously, other we hear the death of a perfect healthy looking person. At times, a person may look healthy but the same person may not be healthy, which gives rise to abnormalities that can affect the outcome of diagnosis.

# 4 Conclusion

In this work, we have upgraded the previously proposed approach for simpler and effective results. We have used transfer learning and logistic regression to provide the desired outcome. We have eliminated the entire idea of the multi-expert approach. We have gotten 99% success rate in the terms of accuracy and effectiveness. Lastly, I would say our approach uses a more refined and simple approach for giving an effective result, which can help thousands of people out in the world.

# References

1. Fernandes, K., Cardoso, J.S., Fernandes, J.: Temporal segmentation of digital colposcopies. In: Pattern Recognition and Image Analysis. Springer, pp 262–271(2015)
2. Pan, S.J., Yang, Q.: A survey on transfer learning. IEEE Trans. Knowl. Data Eng. 22(10), 1345–1359 (2010)
3. Garcke, J., Vanck, T.: Importance weighted inductive transfer learning for regression. In: Machine Learning and Knowledge Discovery in Databases. Springer, pp 466–481 (2014)
4. Rückert, U., Kramer, S.: Kernel-based inductive transfer. In: Machine Learning and Knowledge Discovery in Databases. Springer, pp 220–233 (2008)

# Automatic Examination Timetable Scheduling Using Particle Swarm Optimization and Local Search Algorithm

**Olusola Abayomi-Alli, Adebayo Abayomi-Alli, Sanjay Misra, Robertas Damasevicius and Rytis Maskeliunas**

## 1  Introduction

Educational timetabling challenges include school timetabling, examination timetabling and university course. Basically, university timetable problem exists in two forms, viz course and examination timetable formats [1]. Examination scheduling or timetabling is one of the most difficult problems faced by academic community [2]. Timetabling challenge is defined as assigning examinations into a fixed number of timeslots hereby satisfying all university constraints [3]. However, for every institution the objective is to construct an effective and satisfactory examination timetable. An institutional timetable is said to be effective when it keep its users satisfied to a reasonable extent.

The main challenge of examination timetable scheduling is the population of students when compared to the inadequate available resources within a short period of time; hence, the population of students will always be more than available resources [4]. The problem addressed here deals with the assignment of examinations into a limited number of timeslots with respect to certain constraints for the undergraduate

O. Abayomi-Alli (✉) · S. Misra
Center of ICT/ICE Research, CUCRID Building, Covenant University, Ota, Nigeria
e-mail: olusola.abayomi-alli@covenantuniversity.edu.ng

S. Misra
e-mail: sanjay.misra@covenantuniversity.edu.ng

A. Abayomi-Alli
Department of Computer Science, Federal University of Agriculture, Abeokuta, Nigeria
e-mail: abayomiallia@funaab.edu.ng

R. Damasevicius · R. Maskeliunas
Kaunas University of Technology, Kaunas, Lithuania
e-mail: robertas.damasevicius@ktu.lt

R. Maskeliunas
e-mail: rytis.maskeliunas@ktu.lt

© Springer Nature Singapore Pte Ltd. 2019
R. K. Shukla et al. (eds.), *Data, Engineering and Applications*,
https://doi.org/10.1007/978-981-13-6347-4_11

students with about a population of 15,000 from the Federal University of Agriculture, Abeokuta, Nigeria. There are two major types of constraints to be satisfied when dealing with a university examination timetabling which are: Hard and Soft constraints. Hard constraints are conditions that must be conceited while soft constraints may not be conceited, but it is desirable to have a good and feasible timetable. Hard constraints considered in this study was adopted from [5]: (i) every examination should be assigned to a venue at a particular timeslot; (ii) the scheduled examinations must not exceed the venue capacity; (iii) the maximum number of time period within a day should not be exceeded. Soft constraint include: (i) student preference for having an examination a day; (ii) finalist preference for having their examinations scheduled for first to second week of the examination. Satisfying all the constraints to have a good timetable is becoming increasingly difficult. The motivation of this study is the complexity of satisfying the constraints in examination timetable which includes high time consumption and manpower, thereby being too stressful. However, the major objective of this study is to design an algorithm that performs all hard and soft constraints using optimization algorithm and local search techniques to produce an effective timetable feasible solution. The rest of the paper is organized as follows. The literature review and related work is presented in Sect. 2. Section 3 presents the methodology and the results obtained. Section 4 discusses the implementation of the proposed system and the result obtained. The study concluded in Sect. 5.

## 2   Background and Review of Related Works

This section reviews some related work on examination timetabling, although quite a number of solutions have been suggested on solving course scheduling in a university system. Particle swarm optimization is one of the most common approaches applied by earlier researchers in solving timetabling for institution, metro transit network, etc. [6, 7].

Particle swarm optimization (PSO) is a population-based stochastic optimization mechanism inspired by social behavior of bird flocking or fish schooling. This approach solves a problem by having a population (swarm) of candidate solutions known as particles around in the search space over the particles position and velocity [8]. The velocity directs the flying of potential solution, called particles by following its current position known as local best known position, and is also guided toward the best known position in the search space known as the global best position, tracked by particle swarm optimizer [9]. The main concept of the PSO algorithm comprises of changing the velocity or accelerating each particle toward its gBest and lBest locations at each step. Acceleration is measured by a random term, with separate random numbers being generated for acceleration toward gBest and lBest locations [3]. However, the velocity and position of the particles are updated by Eqs. (1) and (2).

$$V_{(t+1)} = V_{(t)} + c1 * \text{rand}() * (P_{it} - X_{it}) + c2 * \text{rand}() * \left(P_{gt} - X_{it}\right) \qquad (1)$$

$$X_{it} = X_{it} + V_{(t+1)} \qquad (2)$$

where:

$v_{(t)}$ is the velocity component of particle $i$th at iteration $t$;
$v_{(t+1)}$ is the velocity component of particle $i$th at iteration $t + 1$;
$P_{it}$ is best location of particle $i$th and
$P_{gt}$ is the global best position of particle $i$th of the whole swarm,
$c1$ and $c2$ are learning factors for deriving how the $i$th particle approaching the position that is closer to the individual local best and global best,
rand() is a random number in the range [0, 1].


## 2.1 Literature Review

Pillay [10] presented a genetic programming approach for automated induction of construction heuristics for the curriculum-based course timetabling problem on ITC 2007 dataset. Ahmad and Shaari [3] proposed a PSO approach for solving examination timetable generator in three institutions. However, the proposed study has not been implemented and evaluated yet. Ilyas and Iqbal [11] analyzed hybrid heuristic and meta-heuristic technique to solve university course time tabling challenges. The study concluded that hybrid methods require more computational cost and is difficult to implement.

Chen and Shih [4] presented a constriction PSO with local search technique in solving university course timetabling problems. The experimental result shows the improvement in the solution quality and low computational complexity based on the small data sample used for implementation. Oswald and Anand [12] presented a novel hybrid PSO algorithm for solving university course timetabling problem. The experimental result gave a suboptimal solution, with major limitation on the inability to satisfy the soft constraints. Tassopoulos and Beligiannis [13] experimented with dataset from different high schools using PSO algorithm. The result shows that PSO outperformed the other techniques while [8] analyzed the application of PSO in course scheduling system. The study considered the hard and soft constraints and gave a quality solution. Najdpour and Feizi-Derakshi [14], Bhatt and Sahajpal [15], Karol et al. [16] developed an automatic course scheduling system using a generic algorithm, however, the former combined GA with cycle crossover approach for solving soft constraints with an improve solution when compared with the manually generated solution. Karol et al. [16] introduced known parallelization techniques to generic algorithm, which has high computational complexity.

The overall summary of the existing approaches to timetable scheduling problems includes difficulty of incorporating the constraints into scheduling procedures to

**Table 1** Summary of reviewed related work

| Authors | Methodology | Strength | Weakness | Implementation |
|---|---|---|---|---|
| Pillay [10] | Generic programming approach | Reduce soft constraints cost | Computational cost was not stated | Course timetabling using ITC 2007 dataset |
| Ilyas and Iqbal [11] | Hybrid heuristic and meta-heuristic method | Provides better solution | Difficult implementing and increased computational cost | University course timetabling |
| Komaki et al. [17] | Column generation Heuristics | Improved computational time | No realistic constraints was considered | Makeup class timetabling |
| Jaengchuea and Lohpetch [1] | Generic algorithm with local and tabu search approaches | Improved result quality | High computational runtime | Experimented course timetabling |
| Shiau [9] | Particle swarm optimization | Solution with optimal satisfaction | Undefined computational complexity | Course scheduling university in Taiwan |

improve schedule quality. Table 1 shows a summary of some existing solutions on timetabling problems related to this work.

## 3   The Proposed Methodology

This study designed an automatic examination timetable system for the Federal University of Agriculture, Abeokuta (FUNAAB), Nigeria, using particle swam optimization algorithm and local search technique. In FUNAAB, the manual examination timetable was always done by a Timetable Committee known as (TIMTEC). However, in the process of designing the timetable by TIMTEC, several draft copies are corrected before the final timetable is reached due to the manual procedure and complexity of the scheduling; hence, there is a need for an automatic examination timetabling scheduling algorithm. The proposed PSO-based algorithm structure is shown in Fig. 1 while Table 2 shows the pseudocode of algorithm for the PSO-based algorithm, and Table 3 shows the local search techniques algorithm.

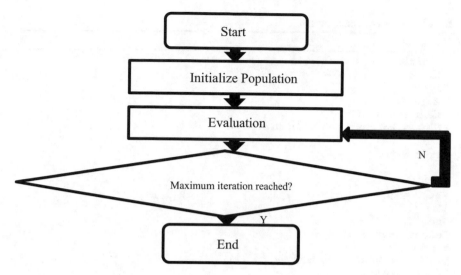

**Fig. 1** Proposed PSO-based algorithm structure

## 3.1 Mapping of University Examination Timetable to PSO

Federal University of Agriculture Abeokuta uses three weeks' time period for her examination, the three weeks duration starts on Monday–Friday due to the general five working days, when it falls during the festive period, then it is likely to start anytime during the week. In FUNAAB, the duration of an examination can be 2, 2.5, or 3 h. The time given on the timetable does not determine the time spent for the examination; the lecturer(s) determines that. Therefore, the preferable length of one timeslot is half an hour.

Considering the theory examination only, we have two sessions; morning (9 am–12 pm) and afternoon (2 pm–5 pm) for the examination. Maximizing the available facilities, examinations can be allocated to venues for 30 time period, considering the availability of venue for each session in the time period (three weeks). The Halls and their capacity are represented using one-dimensional array, respectively, where the name of a particular hall is referencing its capacity using the same index (as shown in Tables 4 and 5). The Examinations and their population will also be stored in one-dimensional array (as shown in Tables 6 and 7).

The total timeslots per day will be 12 with half an hour per timeslot. The day of the timeslot can be calculated by taking the ceiling of the number divided by 12. For example, the day of timeslot number 15 will be equal to the ceiling $(15/12) = 2$ (Tuesday) and the day of timeslot number 38 will be equal to the ceiling $(38/12) = 4$ (Thursday). The whole timeslots of a week can be represented by one-dimensional array as shown in Table 8, the first 12 elements (1–12) of the array will represent the timeslots for Monday, the second 12 elements (13–24) of the array will represent the timeslots of Tuesday, etc.

**Table 2** Pseudocode of algorithm

```
∀ x in X
   Initialize x with r, y, z
End
Do
∀ x in X;
Calculate u of x;
If u is better than v in history
{
Set q = p
}
Else {
Set v = p
}
End
Select best u of X = w
∀ x in X
Calculate y of x according to (1)
Update y of x
Calculate z of x according to (2)
Update z of x
End
While maximum iteration is reached
Where random number = r, population = X, velocity = y, position
= z, fitness value = u, lbest = v, new lbest = p, current value
= q, gbest = w.
```

**Table 3** Local search technique algorithm

```
Input: Hall, Examination
Output: Allocated hall to Exam;
Process:
Step 1: For each examination not yet allocated;
Step 2: Get all halls available for use;
Step 3: Move from hall to hall until a hall can fit for the
population of an exam;
Step 4: Assign hall to exam;
Step 5: Stop search and move to another examination until
exhausted;
Step 6: End.
```

After the PSO algorithm has been applied, Tables 4, 5, and 8 are mapped to PSO to form Table 9. The total number of examinations is represented as one-dimensional

**Table 4** Encoding of halls (H is the number of available halls)

| Index | 1 | 2 | 3 | 4 | ... | H |
|---|---|---|---|---|---|---|
| Hall | Hall 1 | Hall 2 | Hall 3 | Hall 4 | ... | Hall H |

array (Table 4) where the index of the array will represent implicitly the hall name and will also be used to calculate day and session of the day the examination is to hold.

After the allocation of examinations to venues using the PSO algorithm, local search algorithm is applied which works by searching for venues that fit for the population of each examination until examinations are exhausted.

## 3.2 Data Source

In this study, the examination timetable problem at the Federal University of Agriculture, Abeokuta was properly studied. The dataset presented in this study is a real data for 2016 undergraduate examinations for first and second semester. The dataset, the total number of examinations is 889 with about 15,000 students and 39 halls; the capacity of the venues for examinations is different from the normal capacity of venues due to the sitting arrangement during the examination. The number of examination days and timeslots are 15 days with 30 available timeslots. The dataset is available at TIMTEC of the Federal University of Agriculture, Abeokuta.

## 4  Implementation and Results Obtained

This section discusses the implementation and the experimental results obtained from the automatic PSO_LST timetable system. The minimum system requirement for this study includes Window XP or any compatible OS, NetBeans IDE/Eclipse, 2.4 GHz processor, 20 GB Hard disk, and 1 GB RAM. The following are the parameters required to execute this algorithm:

- A list of all courses whose examination is to be done
- A list of all Halls to be used for examination
- The number of particles
- The Constriction (Learning) factors
- The Minimum fitness threshold
- The Maximum fitness threshold

**Table 5** Encoding of hall capacity

| Index | 1 | 2 | 3 | 4 | … | H |
|-------|-----|-------|-------|------|-----|-------|
| Capacity | Cap1 | Cap 2 | Cap 3 | Cap4 | … | Cap H |

**Table 6** Encoding of examinations (N is the total number of examination)

| Index | 1 | 2 | 3 | 4 | … | N |
|-------|-------------|-------------|-------------|-------------|-----|-------------|
| Examinations | Examination1 | Examination2 | Examination3 | Examination4 | …. | ExaminationN |

**Table 7**  Encoding of examination population

| Index | 1 | 2 | ... | N |
|---|---|---|---|---|
| Population | Population1 | Population2 | ... | PopulationN |

**Table 8**  Timeslots of a week in one-dimensional array

| 9:00–9:30 | 9:30–10:00 | 10:00–10:30 | 10:30–11:00 | ... | 4:30–5:00 |
|---|---|---|---|---|---|
| 1 | 2 | 3 | 4 | ... | 12 |
| Monday | | | | | |
| . | | | | | |
| . | | | | | |
| 9:00–9:30 | 9:30–10:00 | 10:00–10:30 | 10:30–11:00 | ... | 4:30–5:00 |
| 1 | 2 | 3 | 4 | ... | 12 |
| Friday | | | | | |

**Table 9**  Mapping of examinations, venues, and timeslots to PSO

| Timeslot | 13 | 14 | 15 | 16 | 17 | 18 | ... |
|---|---|---|---|---|---|---|---|
| Examination | Examination4 | Examination4 | Examination4 | Examination4 | Examination4 | Examination4 | ... |
| Hall | Hall23 | Hall23 | Hall23 | Hall23 | Hall23 | Hall23 | ... |

- Days: the number of days required for the examinations.

Table 10 shows the venue and examination details for FUNAAB displaying the number of sessions per day.

## 4.1  Experimental Results

The algorithm has been tested using the first and second semester examination, session 2015/2016, of the Federal University of Agriculture, Abeokuta. Figure 3 shows the screenshot result from first semester examination scheduling test (Figs. 2, 4, and 5).

**Table 10**  Venue and examination details

| First semester | | Second semester | |
|---|---|---|---|
| Resources | Value | | Value |
| No. of examination | 483 | | 406 |
| No. of halls | 39 | | 39 |
| No. of session | 2 (morning and evening) | | 2 (morning and evening) |

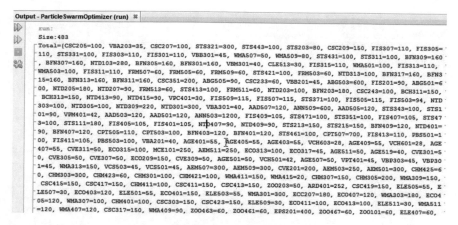

**Fig. 2** Screenshot of examinations and students population for first semester

**Fig. 3** Screenshot of venues and their capacity for first semester

**Fig. 4** Screenshot showing input of parameters for first semester

**Fig. 5** Screenshot showing allocation of examinations to venues for first semester

## 4.2 Results

This section gave an evaluation result of the FUNAAB examination timetable after using the particle swarm optimization technique and a result after combining the PSO and local search algorithm. Table 11 shows the table of the evaluation result obtained.

- Allocation: total number of examinations successfully allocated to an examination hall.
- Un-allocation: total number of examinations not successfully allocated to an examination hall.
- Clashes: Conditions where two examinations are allocated to the same venue.
- Multiple: a condition where a venue has been allocated more than one examination.

**Table 11** Evaluation of results obtained

| S. No. | Condition | First semester Total number of courses = 483 | | Second semester Total number of courses = 408 | |
|---|---|---|---|---|---|
| | | No. of courses | Degree of accuracy (%) | No. of courses | Degree of accuracy (%) |
| 1 | Allocation PSO | 41 | 8.5 | 31 | 7.6 |
| 2 | Allocation PSO-LS | 406 | 84.1 | 311 | 76.2 |
| 3 | Un-allocation | 77 | 15.9 | 95 | 23.3 |
| 4 | Clashes | 49 | 10.2 | 10 | 2.5 |
| 5 | Multiple | 0 | 0 | 0 | 0 |
| 6 | Duplication | 0 | 0 | 0 | 0 |

- Duplication: a condition where an examination has been allocated multiple times.

$$\text{Error}(\%) = (\text{Total number of clashes}/\text{Total number of courses}) * 100 \quad (3)$$

$$\text{Deg. of Acc} = (\text{Total number of allocated}/\text{Total number of courses}) * 100 \quad (4)$$

From the summary of the results on Table 11, using Eqs. (3) and (4), a degree of accuracy of 84%, error of 10% for first semester and a degree of 77%, error of 22% for second semester was observed. Based on this result, PSO algorithm could not provide a perfectly feasible solution but a near-optimal solution with the integration of local search technique as its degree of accuracy is in the range of 77–84%. It is observed from the results that, when the value of the learning factors $c1$ and $c2$ equals 1; more optimal solutions are obtained compare to when $c1$ and $c2$ equals 2.

## 5  Conclusion

In this study, an enhanced particle swarm optimization algorithm was presented for the Federal University of Agriculture, Abeokuta, Nigeria examination timetable scheduling by integrating the local search technique. The integration of the local search technique in the scheduling process helped to increase the effectiveness of the algorithm and the degree of accuracy against using only PSO. From the results obtained, combination of the particle swarm optimization algorithm and local search technique for the University examination timetable approaches near-optimal solution. By using the method of integrating PSO algorithm and LS technique, this research may produce feasible timetable for E-examinations, laboratory, and college timetable. A future research consideration would be to apply the PSO algorithm, LS technique, and course sandwiching to the University course and examination timetable.

**Acknowledgements**  We acknowledge the support and sponsorship provided by Covenant University through the Centre for Research, Innovation and Discovery (CUCRID).

## References

1. Jaengchuea, S., Lohpetch, D.: A hybrid genetic algorithm with local search and tabu search approach for solving the post enrolment based course timetabling problem: outperforming guided search genetic algorithm. In: International Conference on Information Technology and Electrical Engineering, pp. 29–34. IEEE, Chiang Mal (2015)
2. Legierski, W., Widawski, R.: System of automated timetabling. In: International Conference of Information Technology Interfaces (ITI), pp. 495–500. IEEE, Cavtat (2003)

3. Ahmad, A., Shaari, F.: Solving university/polytechnics exam timetable problem. In: 10th International Conference on Ubiquitous Information Management and Communication (IMCOM'16). Association for Computing Machinery, New York (2016)

4. Chen, R.-M., Shih, H.: Solving university course timetabling problems using constriction particle swarm optimization with local search. Algorithms J $6(2)$, 227–244 (2013)

5. Shaker, K., Abdullah, S.: Incorporating great deluge approach with kempe chain neighbourhood structure for curriculum-based course timetabling problems. In: Conference on Data Mining and Optimization, pp. 149–153. IEEE, Selangor (2009)

6. Guo, X., Sun, H., Wu, J., Jin, J., Zhou, J., Gao, Z.: Multiperiod-based timetable optimization for metro transit networks. Elsevier J Transp Res Part B $96$, 46–67 (2017)

7. Aziz, M.A., Taib, M.N., Hussin, N.M.: An improved event selection technique in a modified PSO algorithm to solve class scheduling problems. In: IEEE Symposium on Industrial Electronics and Applications, pp. 203–208. IEEE, Kuala Lumpur (2009)

8. Li, L., Liu, S.-H.: Study of course scheduling based on particle swarm optimization. In: Cross Strait Quad-Regional Radio Science and Wireless Technology Conference (CSQRWC), pp. 1692–1695. IEEE, Harbin (2011)

9. Shiau, D.-F.: A hybrid particle swarm optimization for a university course scheduling. J Expert Syst Appl $38(1)$, 235–248 (2011)

10. Pillay, N.: Evolving construction heuristics for the curriculum based university course timetabling problem. IEEE Congress Evolutionary Computation (CEC), pp. 4437–4443 (2016)

11. Ilyas, R., Iqbal, Z.: Study of hybrid approaches used for university course. In: IEEE 10th Conference on Industrial Electronics and Applications, pp. 696–701, Auckland (2015)

12. Oswald, C., Anand, D.C.: Novel hybrid PSO algorithms with search optimization strategies for a university course timetabling problem. In: 5th International Conference on Advanced Computing (ICoAC), pp. 77–85. IEEE, Chennai (2013)

13. Tassopoulos, I.X., Beligiannis, G.N.: Solving effectively the school timetabling problem using PSO. Expert Syst. Appl. $39(5)$, 6029–6040 (2012)

14. Najdpour, N., Feizi-Derakshi, M.-R.: A two-phase evolutionary algorithm for the university course timetabling problem. In: International Conference on Software Technology and Engineering, pp. 266–271. IEEE, San Juan (2010)

15. Bhatt, V., Sahajpal, R.: Lecture timetabling using hybrid genetic algorithms. In: International Conference on Intelligent Sensing and Information Processing (ICSIP), pp. 29–34. IEEE, Chennai (2009)

16. Karol, B., Tomasz, B., Henry, K.: Parallelization of genetic algorithms for solving university timetabling problems. Parallel Computing in Electrical Engineering. IEEE Computing Society Technical Committee on Parallel Processing (TCPP). IEEE, Great Britain (2006)

17. Komaki, H., Shimazaki S., Sakakibara,K., Matsumoto, T.: Interactive optimization techniques based on a column generation model for timetabling problems of university makeup courses. In: Computational Intelligence and Applications (IWCIA), 2015 IEEE 8th International Workshop on, pp. 127-130. IEEE (2015)

# Personality Trait Identification for Written Texts Using MLNB

S. Arjaria, A. Shrivastav, A. S. Rathore and Vipin Tiwari

## 1 Introduction

Personality of a person is supposed to be the unique set of behavioral characteristics of him, this is can be used to predict the human response in many types of interaction; his attitude toward life, jobs, family, etc. Hence, the written expressions reflect the human's personality, and all the personality are based on three aspects behavior, emotions, mood. The main aim is to define the personality is based on the written texts.

The present era is supposed to be the information age in which the use of social networking has become part and parcel of life. The use of the Internet and social networking sites is used for a variety of purposes covering the personal, social, political, commercial area. These psychological issues are the combined reflection of mood, emotion, and personality of the person. If the Facebook post shared by human is analyzed properly, it will give the approximate estimation of the behavioral-related issues of the individual. A huge number of researchers around the world have been attracted to work on this research domain from different fields, especially researchers in computational linguistics, psychology, artificial intelligence, natural language processing, human–machine interaction, behavioral analytics, and machine learning. Hence, the written expressions reflect the human's personality and all the

S. Arjaria (✉) · A. Shrivastav
Department of CSE, TIT, Bhopal, India
e-mail: arjarias@gmail.com

A. Shrivastav
e-mail: shrivastava.ankita583@gmail.com

A. S. Rathore
SVIIT, SVVV, Indore, India
e-mail: abhishekatujjain@gmail.com

V. Tiwari
Department of CSE, SIRT, Bhopal, India
e-mail: vipintiwari1@gmail.com

© Springer Nature Singapore Pte Ltd. 2019
R. K. Shukla et al. (eds.), *Data, Engineering and Applications*,
https://doi.org/10.1007/978-981-13-6347-4_12

personality are based on three aspects behavior, emotions, mood and where the emotion is the labyrinthine psycho-physiological actions of an individual's position while interacting with other individuals or entities. In most cases, the emotion mining is usually based on the written text. According to Baker [1], the words which we are using in our daily life reflects our personality, thought, thought process. The proposed work is based on the same assumption and identifies a person uniqueness from written text with the help of Big Five trait taxonomy [2]. The need of a consistent method and improved performance to analyze the human personality based on the written texts, that easily describe personality of individual, led many researches to efforts to categorize personality with distinct aspects.

The personality prediction may be applied in different domains, like, job recruitment, analyzing deception, friends, matrimonial, and social Web sites [3]. It might be useful for developing user-friendly systems that are customized as per personality of end user with supervised learning. The proposed work identifies the personality on an individual using Big Five Personality traits [2] model from a text corpus tagged for the same purpose. Information extracted from the dataset is purely linguistic. The written texts and Web blogs and different social media platforms are used to train the system when no training data is available. The prediction of personality based on the texts and blog written texts. The written texts are the most important aspect to understand the behavior and personality. The proposed work is intended to provide the improved techniques to predict the personality with machine learning technique.

## 2 Literature Survey

The literature review is all about the personality prediction task from various journals and research papers. Many methods are proposed using different approaches to the solution of the prediction problem and techniques are executed to provide personality prediction. The research work done so far mainly focuses on the written text for predicting personality. Marissa et al. [4] predict the personality through linguistic speaker of the automatic recognition in conversation of text with LIWC. The linguistic provides various information including emotions, etc., as discussed earlier.

Wei et al. [5] predict the personality prediction using Web-searching results. The text search results are categorized with the multi-label learning using Naive Bayesian (NB) multi-label classification algorithm.

Saxena et al. [6] identify the human traits from blog data. The feature vector matrix filtered the emotional intelligence and support vector machine is used for classification.

Sumner et al. [7] proposed dark triad personality traits for linguistic analysis from Twitter usage. The predictive analysis masks inaccuracies the top and bottom percentiles prediction.

Farnadi et al. [8] recognize personality traits using Facebook status for personalization, like automatic recommendation systems and advertising.

Agrawal et al. [9] used the Big Five model to predict personality detection from text. They used a Big Five model. It is used as a personality detection model to extract correlation patterns between personality and variety of data captured from human traits. Two types of techniques that have been employed for the detection of personality from text and machine learning approach based on social network activities and the second is based on linguistic properties based on the text. Use of social networking has increased tremendously in recent times. It has become a popular method for information distribution and social interaction. Personality has been considered as the most difficult human attribute to understand. It is very important as it can be used to define the uniqueness of a person. Personality detection from text means to extract the behavior characteristics of authors written the text.

## 2.1 The Big Five Model

The Big Five model of personality dimensions is one of the most emerging areas for researcher to personality identification in recent years. The Big Five model is taxonomy of personality traits, maps which traits go together in people's descriptions of ratings of one another. Factor analysis, a statistical method is used to analyze different personality traits, is employed to discover the Big Five factors [2]. Today, many researchers are agreed on the five core models of personality, openness, conscientiousness, extroversion, agreeableness, neuroticism (Fig. 1).

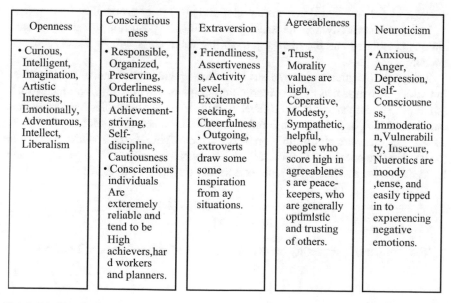

Fig. 1  Big Five model [2]

## 3   Proposed Work

The proposed work processed the essay datasets for the identification of the personality traits using MLNB classifier. The proposed framework has been developed in modules, namely preprocessing, learning, classification, and evaluation. The task of predicting user's personality through the written text is complex task. In this work, the first essay dataset has been explored then the preprocessing step has been carried out that comprises the technique and processes to make the texts data suitable for classification purpose. In preprocessing stage is formulated by the stemming, stops word removal, and normalized the written datasets and these datasets are given with a frequency values and it is represented as the vector space model. Finally, the estimating results using MLNB algorithm to identify the personality traits [OCEAN] extracted from the human writing assignment.

### 3.1   Multi-label Naïve Bayes (MLNB) Algorithm

A multi-label Naïve Bayes classifier provides relatively better performance on classification tasks. It assumes that features are independent and generated randomly. A Naïve Byes classifier is defined as

$$Fi(x) = \prod p(xj|ci)$$

where $X = (x1, x2, \ldots xn)$ is feature vector and $p(ci)$ prior probability.

In order to correctly find the class of test vector, the posterior probability of each class is calculated and the highest probability value class will be the class of test vector.

The pool of written text comprises $D$ document and each document $dj$ is defined in vector space model as the set of terms and frequency is $(a1; a2; \ldots am)$, here $ak$ is $k$th term in document $dj$. In multi-label classification, the examples have more than one target class, and thus, each training sample has a set of disjoint labels $Y \subseteq L$, $|L| > 1$. NB classifier algorithm is applied for identifying the personality traits from the written document $d \in D$ for it $|L| = 5$ binary NB classifiers $Hl: d \rightarrow \{l; \neg l\}$ are learned, one for each different label personality trait $\{O, C, E, A, N\}$

$$H(d) = \bigcup_{l \in \{O,C,E,A,N\}} \{l : H(D) = I\}$$

The number of personality traits that a document can be between 1 and 5.

# 4  Result and Discussion

The essay dataset is same as used in [1] for evaluation of the proposed work. It contains 2468 text documents. The dataset was generated by asking students to write randomly whatever they think for 20 min. These documents were tagged for the Big Five Personality Traits, recognized in one topic.

To evaluate the system performance, accuracy measure is used. The algorithm is trained with 10fold cross-validation. The original texts contain 2400 distinct words. The evaluation measure is performed using personality traits and best accuracy result openness, conscientiousness, extraversion, agreeableness, neuroticism.

## 4.1  Result Through MLNB with Different Tested Data

The experiment is executed with MLNB and different tested data are evaluated which written texts are formed which type of traits are there (Table 1).

## 4.2  Comparative Analysis of Proposed Work with Existing Work

This is the comparative result analysis of personality prediction [10] with written datasets with the learners evolution algorithm. The different test data are measured with both the algorithms and observing that the MLNB algorithm has performed better prediction than the others (Table 2).

**Table 1**  Result analysis on essay dataset through MLNB

| Tested features | Personality traits | | | | |
|---|---|---|---|---|---|
| | Openness | Conscientious | Extraversion | Agreeableness | Neuroticism |
| 100 | 98.3281 | 77.2180 | 77.8081 | 85.0109 | 80.0420 |
| 200 | 94.1484 | 79.8215 | 79.8765 | 96.7880 | 65.8403 |
| 300 | 99.8955 | 64.1731 | 53.5354 | 53.4047 | 71.2185 |
| 400 | 98.5371 | 96.3865 | 69.4949 | 41.2634 | 74.4748 |
| 500 | 96.1787 | 76.7603 | 73.7374 | 97.7516 | 13.6555 |

**Table 2** Comparative study

| Algorithms | Personality traits | | | | |
|---|---|---|---|---|---|
| | Openness | Conscientious | Extraversion | Agreeableness | Neuroticism |
| MLNB | 99.8955 | 96.3865 | 79.8765 | 96.7880 | 65.8403 |
| Learnable evolution | 83.07 | 77.73 | 63.07 | 84.67 | 54.53 |

## 5 Conclusion

The research is focused to improve performance automatic personality prediction. Many criteria such as written text, online status updates, Facebook posts, twitter texts are available to judge the personality. In this work, written texts' are used to predict and define the personality criteria. People written their texts and our aim is to extract the features encapsulate through various techniques to investigate what type of written texts is which type of traits are there. There are various advantages of personality prediction such as business intelligence, human resource employees, marketing to understand the customer the behavior, advertisement team, which type of brand preferred liked the most, that is to understand these to which determined the personality of humans. The study of MLNB method has been performed and results are discussed.

## References

1. François, M., Walker, A., Matthias, R.M., Roger, K.: Moore: using linguistic cues for the automatic recognition of personality in conversation and text. J. Artif. Int. Res. **30**(1), 457–500 (2007)
2. Barrick, M.R., Mount, M.K.: The Big Five personality dimensions and job performance: a meta-analysis. Pers. Psychol. **44**, 1–26 (1991)
3. Golbeck, J., Robles, C., Turner, K.: Predicting personality with social media. In: CHI'11 Extended Abstracts on Human Factors in Computing Systems, pp. 253–262. ACM, Vancouver (2011)
4. Marissa, F., Walker, M.A., Mehl, M.R., Moore, R.K.: Using linguistic cues for the automatic recognition of personality in conversation and text. J. Artif. Intell. Res. **30**, 457–500 (2007)
5. Wei, Z.: A naive Bayesian multi-label classification algorithm with application to visualize web search results. Int. J. Adv. Intell. **3**, 173–188 (2011)
6. Saxena, S., Sharma, C.M.: Machine learning based approach for human trait identification from blog data. Int. J. Comput. Appl. **48**, 17–22 (2012)
7. Sumner, C., Byers, A., Boochever, R., Park, G.J.: Predicting dark triad personality traits from twitter usage and a linguistic analysis of tweets. In: International Conference on Machine Learning and Applications, pp. 386–393, IEEE, Boca Raton (2012)
8. Farnadi, G., Zoghbi, S., Moens, M.F., DeCock, M.: Recognising personality traits using face book status updates. Technical report, AAAI (2013)

9. Agrawal, B.: Personality detection from text. Int. J. Comput. Syst. **01**, 1–4 (2014)
10. Gupta, U., Chatterjee, N.: Personality traits identification using rough sets based machine learning. In: International Symposium on Computational and Business Intelligence, pp. 182–185. IEEE, New Delhi (2013)

# Deep Neural Network Compression via Knowledge Distillation for Embedded Vision Applications

Bhavesh Jaiswal and Nagendra P. Gajjar

## 1 Introduction

Since the last half decade, deep learning has pushed the boundaries of the machines to imitate the human-like decision-making capabilities. Graphics Processing Units (GPUs) collated with neural architectures are the major contributors for this development. GPUs with their unique parallel processing of the data chunks have overshadowed the other research areas in this field. Knowledge distillation is one such idea initially proposed by Bucila et al. [1] and taken afresh with a new perspective in 2015 by Hinton et al. [2]. In this approach, the lightweight model, called the student, learns from a high-level model, called the teacher. The teacher–student approach tries to imitate the transfer of knowledge from one entity to other. The high-level teacher is a heavyweight model, regarding memory size, runtime memory requirement, computation cost, computation time, etc. This learning approach is feasible provided a high-end GPU, which can do the computation in the limited period, as the main processor alone is not capable of finishing these computations in the same time duration.

In this work, a teacher–student model is proposed for applications having lesser computation capabilities. The student is trained along with the teacher, and the neural networks can be deployed in memory-limited settings. The proposed structural model distillation for memory reduction architecture is explored using a strategy to have a student model that is a simplified teacher model: no redesign is needed, and the same hyperparameters can be used. With this approach, there are substantial memory savings possible with very little loss of accuracy and the knowledge distillation provides the student model to perform better than training the same student model directly on data.

B. Jaiswal · N. P. Gajjar (✉)
Electronics and Communication Engineering Department, Institute of Technology,
Nirma University, Ahmedabad, Gujarat, India
e-mail: nagendra.gajjar@nirmauni.ac.in

© Springer Nature Singapore Pte Ltd. 2019
R. K. Shukla et al. (eds.), *Data, Engineering and Applications*,
https://doi.org/10.1007/978-981-13-6347-4_13

## 2   State of the Art

Figure 1 below demonstrates an architecture of a typical network of neurons receiving data from the external world into the input layer and gives classification results as class probabilities through the output layer. There are multiple interconnected hidden layers between the input and output layer. These hidden layers could have feed-forward or feedback connections. All the layers consist of a basic unit called neuron (shown as circles in Fig. 1). The computational output of one neuron is passed on to the neurons in the next layers.

The neurons are triggered nonlinear activation function, like biological neurons in our brain. The present work considers feed-forward-type connection in which information is transferred from the input side to the output side. Neural network training involves the weights initialisation and updates after every iteration. The network weights/hyperparameters are normally frozen before deploying the model for prediction task. To reduce the training time as well as prediction time, the various model compression methods are available which can be categorised as below:

- Network Pruning
- Network Compaction
- Parameter Sharing
- Transfer Learning

### 2.1   Network Pruning

Deep neural networks have millions of hyperparameters which make training difficult to implement in the resource-constrained embedded application. A vast amount of these are redundant, so the network is pruned to reduce these hyperparameters. Network fine-tuning, from the pre-trained weights and reduced hyperparameters set,

**Fig. 1**   An architecture of a typical network of neurons

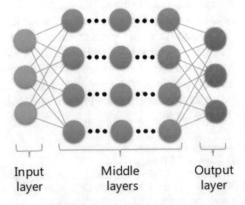

Input        Middle        Output
layer        layers        layer

is required to maintain the accuracy level at par with the original model. The criteria for pruning can be implemented with different parameters.

In [3], the weights with the smaller magnitudes are removed, and the model is then fine-tuned to get the accuracy of over 80%. In work [4], the weight pruning is directly controlled by energy distribution for the network. Energy Inference Engine (EIE) [5] uses sparse matrix-vector multiplication for weights reduction. In this technique, when the input is not zero, the weights (stored in run-length format) are updated. Huffman coding is further used to compress the sparse weights which reduce bandwidth and storage requirements for pruned weights by 20–30% [6].

## 2.2 Network Compaction

Instead of pruning the network, the architecture can be improved itself by using various filter size reduction approaches. The trend in vision-based applications is to replace a large convolution operation with the combination of the cheaper convolutions having fewer weights in total, so that we have the same active filter region from the input image to compute an output. These smaller filters can be concatenated to emulate a larger size filter. In [7], one 'N × N' convolution is decomposed into one '1 × N' and one 'N × 1' convolution. As demonstrated in [8, 9], the number of inputs channel to the next '1 × 1' convolutional layers can reduce the required number of filters for the next layers.

SqueezeNet [10] first uses '1 × 1' convolution filter aggressively and then it expands that with the multiple '1 × 1' and '3 × 3' convolution types. By doing this, it gets a reduction in weights by 50× compared to original AlexNet, while getting the reasonable level of accuracy at par with the original model.

## 2.3 Parameter Sharing

Another approach for reducing parameters computation cost is to share them among the layers. Hashed Nets model was proposed in work [11] on these lines, which group weights into hash buckets. 'k-means' clustering is used for quantisation of the weights in fully connected layers and achieved up to 24× compression rate for the convolutional neural network with the same accuracy on the ImageNet challenge [12]. In the [1], authors used a regularisation technique, instead to quantising the weights in fully connected (FC) layers, to improve model prediction.

## 2.4  Transfer Learning

Transfer learning strategy is recently gaining momentum due to its usability to transfer the knowledge gained from one network to another network ensemble with the same or new domain of application for the network. One of the transfer learning approach, used in this work, adopt a teacher–student strategy, where a large complex network trained for a particular task teaches lesser complex student network on the same task. This learning method is initially proposed in [13] in which a synthetic data, containing the knowledge of the teacher model, was created by labelling the unlabelled data using the teacher model. Then, a smaller student model is trained. The next section elaborates the concept of knowledge distillation as one of the techniques for transfer learning.

## 3  Knowledge Distillation

In [2], Ba and Caruana initially proposed the student network model training by replicating the *logit* values (the softened class probabilities) of the teacher network. Their work is then extended by using intermediate layer outputs as the soft target values for the training of the student network/model [14]. This can be generalised by introducing a temperature variable in the softmax function [15]. They showed that the softened outputs at higher temperatures convey important information. They termed this information, which is expressed by the relative scores (probabilities) of the output classes as 'dark knowledge'. Deep neural network compression can be categorised depending on the following three parameters:

- Training cost
- Runtime requirements
- Storage requirements

Network pruning has a good impact on the initial and runtime storage requirements but is not aimed at reducing the other runtime computation complexities, whereas network compaction reduces convolution/filter size and thus the weights to be stored or passed on. Reduction in the number of weights does not mean a reduction in energy consumption, as SqueezeNet runtime utilises more energy than AlexNet [16]. Parameter sharing methods focus on only the storage complexity of the deep models but again fail to improve on runtime complexity. Knowledge distillation method can achieve compression across all the three perspectives, but there has not been much follow-up work since it was proposed. This observation, supported by the potential of such methods in addressing the challenges above, motivates our proposed work.

Figure 2 presents the concept of knowledge distillation, adapted from [10]. In the image classification task by a DNN, the softmax layer squashes the class scores (*logits*) into the range [0, 1] and gives the class probabilities. As the small-valued scores contain the network/weight information which becomes almost zero after

**Fig. 2** Concept of
knowledge distillation

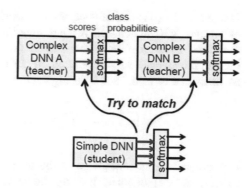

squashing, these are used for knowledge distillation task. The student tries to match
with the teacher network by minimising the squared difference between logits of
teacher network and a class score of the student network. In another approach, soft
class probabilities could be generated from a softmax layer to compare the class
probabilities of the two networks directly.

In teacher–student frameworks in deep learning, the teacher is a pre-trained deep
model, which is used to train another (typically shallow) model, called the student.
There has been limited earlier work so far in this context, as described earlier. However, there are significant advantages of using a teacher–student framework beyond
just model compression as observed by [14].

- The 'dark knowledge' present in the teacher outputs works as a powerful target-
  cum-regularise for the student model, as it provides soft targets that share helpful
  information.
- Convergence is typically faster than using only original 0/1 hard labels, due to the
  soft targets, that help training.
- A little amount of training set is generally sufficient for the student network.

## 4 Implementation Choices

Several open-/closed source deep learning frameworks have been developed from
various academic and industry sources. The purpose of the open-source work is to
enable the sharing of the trained networks for the ease of further improvements in
them. The closed source development also releases APIs for integration into existing frameworks. Pre-trained models can be sourced from various websites [17–20].
*Caffe* from UC Berkeley [15] is one such framework which supports Python/C and
MATLAB-based implementations. Google's *Tensorflow* [20] supports Python/C++
programming and mobile CPUs as well as GPUs. It has more flexibility than Caffe, as
the computation here is expressed as data flow graphs to handle the multidimensional
arrays called *'tensors'*. *Torch* developed jointly by NYU and Facebook, supporting

Lua and C/C++, is another popular framework. Other mostly used frameworks such as Theano, MXNet and CNTK are outlined in [21]. It should be noted that even for the same implementation of the network (say, AlexNet) the accuracy of the implemented models can vary depending on the model training parameters. Any model accuracy can be calculated by using the log-loss formula:

$$\text{Logloss} = -\frac{1}{n} \sum_{i=1}^{n} \left[ y_i \log(\hat{y}_i) + (1 - y_i) \log(1 - \hat{y}_i) \right]$$

where $n$ is the number of images in the test set, $\hat{y}_i$ is the prediction probability for the dog. '$y_i$' is '1' if the image is a dog and '0' if a cat. A smaller log loss is better [22].

For the work presented here, the Caffe framework is used because of its wide acceptability and ease of use with Python environment. The log-loss probability is calculated for each run for comparison purpose. The set-up and results are elaborated in the next section.

## 5   Experimental Set-up and Evaluation

For the experimental set-up and evaluation purpose, the Caffe framework is used for image classification task on Redux Dogs versus Cats competition dataset [22] which is divided into two sets—the training set and the testing set. The model outputs a prediction score by calculating the probable class for the pet. If more than two class of pets need to be classified, the prediction score gives the value of probability of each class, then the highest probability class is considered for the pet.

During the training/learning part, the higher complexity teacher model uses labelled data from the training dataset. The learned teacher model is then tested with the unlabelled dataset to classify the pet (cat or dog). Figure 3 shows the training curve of the original model-1. It shows that the teacher model achieves 75% validation accuracy, and saturates in 1500 iterations.

In the second part of the run on model-2, the weights from the model-1 are carried forward to the student model-2 during the learning phase. The learning curve in Fig. 4 shows that the model achieves a 95% validation accuracy within 1000 training iterations. Comparing with Fig. 3, it is quite clear that the student model-2 achieves the similar accuracy within a lesser number of learning iterations. So, both the criteria of faster learning as well as no loss of accuracy are satisfied.

From Table 1, one can deduce that learning accuracy has increased from 75 to 95% with lesser optimisation time/resources consumed. Also, the log loss has been brought to the reasonable limits. The ideal log loss for this particular problem nears to the value one but should be less than 1. This table is drawn for the run on NVIDIA GTX 770 having 1536 cores [23].

**Table 1** Comparison of performance for the two models used in the experiment

| Parameters | Model-1 | Model-2 |
|---|---|---|
| Learning curve | 75% validation accuracy @ 1500 iterations | 95% validation accuracy @ 1000 iterations |
| Solver time | 2 h and 32 min | 1 h and 49 min |
| Log loss (as per equation) | 8.96 | 1.10 |

**Table 2** Comparison of the models used in the experimental set-up

| Platform used | Model-1 (Training run) | | Model-2 (Transfer learning) | |
|---|---|---|---|---|
| NVIDIA GPU type | Accuracy | Number of iterations | Accuracy | Number of iterations |
| GTX 770 | 75 | 1500 | 95 | 1000 |
| GTX 1080Ti | 75 | 1000 | 96 | 1000 |
| | 90 | 4000 | | |
| Dual GTX 1080Ti | 75 | 1000 | 96 | <1000 |
| | 90 | 3000 | | |

The result for a run on NVIDIA GTX 1080Ti having 3584 cores with single GPU, and Dual GPU are shown in Figs. 5 and 6, respectively, and all runs are compared in Table 2.

Table 2 shows the comparison matrix from all the test runs on various GPU platforms regarding accuracy achieved in the number of iterations during the learning

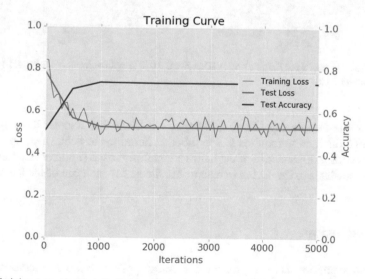

**Fig. 3** Training curve for the original model-1

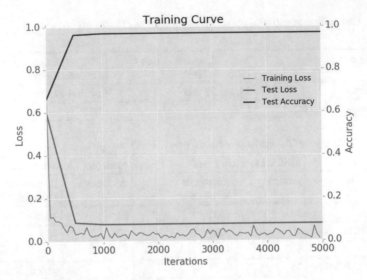

**Fig. 4** Training curve for the student model-2

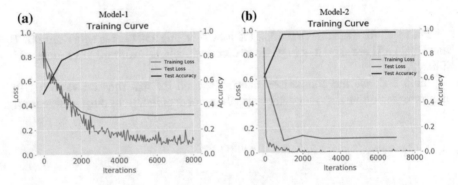

**Fig. 5** Training curve for the run on NVIDIA GPU 1080Ti with same teacher model-1 (**a**) and student model-2 (**b**)

period of the two models. From this table, it is clear that transfer-learning-based knowledge distillation requires less number of iterations to achieve comparable or higher accuracy, irrespective of the hardware platform used (number of cores). This directly translates to the saving in resource and cost for each run of the learning.

# 6 Conclusion

The main requirements for any machine learning application are the prediction accuracy and inference speed. These setting can be easily satisfied with the high-end

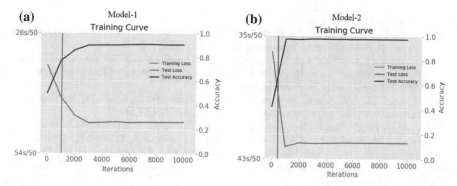

**Fig. 6** Training curve for the run on NVIDIA GPU Dual-1080Ti with same teacher model-1 (**a**) and student model-2 (**b**)

GPU, and a lot of research is going on to optimise for the various use-case scenarios. After training a large, deep model, it may be prohibitively time-consuming to design a model compression strategy to deploy it. On many problems, it may also be more difficult to achieve the desired performance with a smaller model. This model compression strategy using learning transfer is fast to apply and does not require any additional engineering for image classification task. Furthermore, the optimisation algorithm of the larger model is sufficient to train the cheaper student model which can be deployed on various embedded platforms like mobiles or general-purpose experimentation platforms like raspberry-pi.

This work presents the transfer learning methodology, so that DNN can be used on embedded vision-based applications. The trained model knowledge (learned weights) is transferred to another lighter model to learn/infer the vision-based task quickly. An FPGA-based implementation of the approach is carried out to introduce the adaptability to the network.

**Acknowledgements** The authors are extremely thankful to Institute of Technology, Nirma University, Gujrat, in supporting the experimental work. Most of the experiments presented in this paper were carried out using the Grid5000 testbed [23].

# References

1. Bucila, C., Caruana, R., Niculescu-Mizil, A.: Model compression. In: Proceedings of the 12th ACM SIGKDD International Conference on Knowledge Discovery and Data Mining, pp. 535–541. ACM (2006)
2. Hinton, G., Vinyals, O., Dean, J.: Distilling the knowledge in a neural network. arXiv preprint. arXiv:1503.02531 (2015)
3. Han, S., Pool, J., Tran, J., Dally, W.J.: Learning both weights and connections for efficient neural network. In: Advances in Neural Information Processing Systems pp. 1135–1143, (2015)
4. Yang, T.-J., Chen, Y.-H, Sze, V.: Designing energy-efficient convolutional neural networks using energy-aware pruning. In: IEEE Conference on Computer Vision and Pattern Recognition

(CVPR) (2017)

5. Han, S., Liu, X., Mao, H., Pu, J., Pedram, A., Horowitz, M.A., Dally, W.J.: EIE: efficient inference engine on compressed deep neural network, in ISCA (2016)
6. Szegedy, C., Vanhoucke, V., Ioffe, S., Shlens, J., Wojna, Z.: Rethinking the inception architecture for computer vision. In: Proceedings of the IEEE Conference on Computer Vision and Pattern Recognition, pp. 2818–2826 (2016)
7. Szegedy, C., Liu, W., Jia, Y., Sermanet, P., Reed, S., Anguelov, D., Erhan, D., Vanhoucke, V., Rabinovich, A.: Going Deeper With Convolutions, in CVPR (2015)
8. Lin, M., Chen, Q., Yan, S.: Network in network. arXiv preprint. arXiv:1312.4400 (2013)
9. Iandola, F.N., Moskewicz, M.W., Ashraf, K., Han, S., Dally, W.J., Keutzer, K.: SqueezeNet: AlexNet-level accuracy with 50x fewer parameters and <1 MB model size, ICLR (2017)
10. Sze, V., et al.: Efficient processing of deep neural networks: a tutorial and survey. arXiv preprint arXiv:1703.09039 (2017)
11. Gong, Y., Liu, L., Yang, M., Bourdev, L.: Compressing deep convolutional networks using vector quantization. arXiv preprint. arXiv:1412.6115 (2014)
12. Souli´e, G., Gripon, V., Robert, M.: Compression of deep neural networks on the fly. arXiv preprint. arXiv:1509.08745 (2015)
13. Ba, J., Caruana, R.: Do deep nets really need to be deep? In: Advances in neural information processing systems, pp. 2654–2662 (2014)
14. Romero, A., Ballas, N., Kahou, S.E., Chassang, A., Gatta, C., Bengio, Y.: Fitnets: hints for thin deep nets. arXiv preprint. arXiv:1412.6550 (2014)
15. Jia, Y., Shelhamer, E., Donahue, J., Karayev, S., Long, J., Girshick, R., Guadarrama, S., Darrell, T.: Caffe: convolutional architecture for fast feature embedding. In: Proceedings of the 22nd ACM International Conference on Multimedia. ACM, pp. 675–678 (2014)
16. Chen, W., Wilson, J.T., Tyree, S., Weinberger, K.Q., Chen, Y.: Compressing neural networks with the hashing trick. CoRR, abs/1504.04788 (2015)
17. Caffe LeNet MNIST. http://caffe.berkeleyvision.org/gathered/examples/mnist.html
18. Caffe Model Zoo. http://caffe.berkeleyvision.org/modelzoo.html
19. Matconvnet Pretrained Models. http://www.vlfeat.org/matconvnet/pretrained/
20. TensorFlow-Slim Image Classification Library. https://github.com/tensorflow/models/tree/master/slim
21. Deep Learning Frameworks. https://developer.nvidia.com/ deep-learning-frameworks
22. Kaggle's Dogs Versus Cats Competition. https://www.kaggle.com/c/dogs-vs-cats
23. Grid 5000. https://www.grid5000.fr

# An Elective Course Decision Support System Using Decision Tree and Fuzzy Logic

Sushmita Subramani, Sujitha Jose, Tanisha Rajesh Baadkar
and Srinivasa Murthy

## 1 Introduction

Nowadays, many universities offer their students, elective courses as part of the undergraduate and graduate curriculum. For example, the outgoing batch of 2017 at PESIT were offered three electives in their 6th and 7th semesters, and each elective had a choice of six courses. Electives are attractive to students since it offers them a choice of courses. Usually, an elective is chosen considering factors like interest of the student, career path, field of study, student aptitude, etc. Detailed information on career path and field of study (domain) is available in the course handbook and in public domain (Internet). The difficulty level of the elective and how the student may perform on that elective is not easily apparent. There is a lack of student advisory system, particularly in the undergraduate level, to guide the student into the right path about the elective courses that they should take. The student's decision is usually influenced by his/her peers and/or seniors who may have different interests and aptitudes. Choosing the right elective courses is important for the student since it has both short-term and long-term benefits. In the short term, the elective performance would impact the students overall grade point average. Final year coursework and project work will also depend on the student's performance on these electives. In the long run, this will influence his/her career path and specialization. Information about how other similar students have performed in these electives is very helpful to the

S. Subramani · S. Jose (✉) · T. R. Baadkar · S. Murthy
PES University, 100 Feet Ring Road, Banashankari Stage III, Bengaluru 560085, India
e-mail: sujithajose12@gmail.com

S. Subramani
e-mail: Sushmitasubramani@gmail.com

T. R. Baadkar
e-mail: tanisha.baadkar@gmail.com

S. Murthy
e-mail: hvsrinivasamurthy@pes.edu

© Springer Nature Singapore Pte Ltd. 2019
R. K. Shukla et al. (eds.), *Data, Engineering and Applications*,
https://doi.org/10.1007/978-981-13-6347-4_14

149

student to make the choice. Analyzing the subjects that the student has already taken and completed successfully, it is possible to determine the student's aptitude and potential. Electives and the corresponding prerequisite courses taken by past student and their performance can be analyzed, and rules can be mined or extracted.

Decision Tree is a good classification tool since it is simple, easy to understand and performs reasonably well. We use one of the classic decision tree model C4.5, to predict the student elective grade based on the performance of prerequisites. Usually, the prerequisites are foundation courses containing subject knowledge necessary for good elective performance. We have assumed that the performance of the prerequisites determines the most probable grade in the elective.

In addition, the student may also be interested in learning about his potential for success when taking the elective. So, a prediction of "HIGH" performance may be more useful than a grade prediction of "B." Further, students who scored marks in prerequisites near the grade cutoff points are not fully characterized by a single grade. So, we also build a fuzzy logic system that can model this performance uncertainty. We then compare the performance of these two systems.

## 2   Related Work

There is considerable research regarding course recommendation systems (CRS) where variety of data mining methods and approaches are used. These systems assist the student in selecting one or more courses depending on student profile and past history. Bendakir and Aimeur [1] present a CRS called RARE that uses association rules. From historical data, it mines significant rules about former student experience and combines with current student ratings to recommend the most relevant courses. For student course planning, Golumbic [2] has developed a knowledge-based expert system to schedule desired courses for degree completion. It takes into account preferences, strengths of the student, and the remaining degree requirements, to advise prerequisites and exemptions and also schedule the desired courses. Kristiansen et al. [3] show that mathematical models like integer programming and branch-and-price framework can be used for a better planning of elective courses. In [4], Parameswaran et al. study the problem of making course recommendations which satisfy certain requirements. Their system recommends courses that not only satisfy constraints but are also desirable. The models developed are quite expressive and check if the requirements are satisfied. Course recommendations are increasingly used in online learning environment, where the students do not have a college environment and access to peers and former students. Farzan and Brusilovsky describe CourseAgent [5] which is an adaptive, interactive community-based hypermedia system. It provides course recommendations based on students' assessment of course relevance to their career goals. It obtains explicit feedback from the students, which is then used for evaluation. Klasnja-Milicevic et al. [6] present a personalized e-learning system that fits the goals, interests, and skills of the students. This adaptive system recognizes different learning styles, habits of the learners through online monitoring, and mining server logs. After clustering and mining frequent sequences by Aprio-riAll algorithm, it then provides personalized recommendation. Aher and Lobo [7] combine clustering and association rules to recommend course to the student based on the previous choice of other students. This approach of recommending courses to new students is very useful in "MOOC (Massively Open Online Courses)."

Sobecki and Tomczak [8] use ant colony optimization (ACO) to develop an effective course recommendation system. It can predict the final grades of the student using integrated hybrid approach information filtering based on ACO. It is also possible to suggest specialization based on student course history. Harsiti et al. developed a hybrid model from fuzzy Mamdani and C4.5 algorithm to determine the major specialization in informatics engineering courses at Raya Serang Universities [9].

Adak et al. [10] have developed a system to suggest elective courses by analyzing transcripts of students and generating rules from decision tree. The system uses fuzzy logic to suggest electives. They were able to conclude that performances of students in required courses reflect its performance in the related elective courses.

## 3 Methodology

### 3.1 C4.5 Classification

Decision tree (DT) has for decades proved itself to be the most effective method of dividing data set into groups as consistent and uniform as possible, in terms of the number of variables to be predicted. C4.5 is an extension of the basic ID3 DT algorithm designed by Quinlan [11, 12] to address issues such as overfitting data and handling continuous attributes. The DT starts from the root node and constructed using a top-down approach, recursively splitting the data set as the tree is built. At each node, C4.5 considers all possible tests and selects that attribute test which has the highest information gain (or highest gain ratio). The data set is partitioned, and the branches are grown corresponding to the test outcomes. For each discrete attribute, there will be as many outcomes as the number of values. For continuous attribute, data is first sorted and the gain is calculated based on binary cuts of attribute value. The split point can also be midpoint between two adjacent attribute values. This process is repeated for all attributes. The C4.5 algorithm prunes completed tree, which increases the accuracy of test data. C4.5 handles large number of value sets for each attribute, missing attribute values and has good performance [13]. It also avoids creating a wide decision tree.

Weka is a collection of machine learning algorithms for data analytics and data mining tasks [14]. The algorithms can either be applied directly using Weka explorer interface or can be programmatically called using a set of libraries. Weka contains tools for data preprocessing, classification, regression, clustering, association rules, and visualization. We have used Weka to construct the decision tree using its implementation of C4.5 (called J48) and then to extract rules. An attribute-relation file format (ARFF) file is an ASCII text file that describes a list of instances sharing a set of attributes. The header of the ARFF file contains the name of the relation, a list of the attributes (the columns in the data) followed by the data section that contains the actual data.

The synthetic minority oversampling (SMOTE) is a technique used to increase the minority instances in the data set thereby increasing the sensitivity of a classifier/decision tree to the minority instances. We have used the SMOTE algorithm to increase the minority classes in the data set, since the classes in the original data set (student grades) were not balanced.

## 3.2 Fuzzy Logic

A fuzzy logic system (FLS) can be defined as the nonlinear mapping of an input data set to a scalar output data [15]. A FLS consists of four main parts: fuzzifier, rules, inference engine, and defuzzifier. These components and the general architecture of a FLS are shown in Fig. 1. Linguistic variables are the input or output variables of the system whose values are words instead of numerical values. A linguistic variable is generally decomposed into a set of linguistic terms. Membership functions are used in the fuzzification and defuzzification steps of a FLS, to map the non-fuzzy input values to fuzzy linguistic terms and vice versa. A membership function is used to quantify a linguistic term. For example, Table 2 and Fig. 2 show the membership function to map the grade points to grade level.

The FLS process can be described by the following algorithm:

1. Identify variables and define the linguistic terms.
2. Construct the membership functions.
3. Create the set of rules.
4. Fuzzify the input data
5. Evaluate and select the rules in the rule base.
6. Combine the results of each rule.
7. Defuzzify the output data

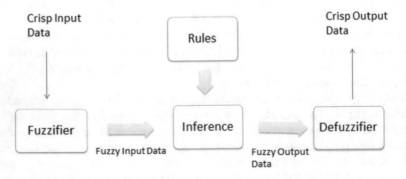

**Fig. 1** Fuzzy logic system

**Fig. 2** Fuzzy membership function

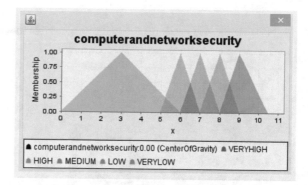

jFuzzyLogic [16] is an open source Java library that offers a fully functional and complete implementation of fuzzy inference system. It provides a programming interface that we have used in the E-DSS. The Fuzzy Control Language (FCL) is a standard for implementing fuzzy logic and serves as an input file. It contains information about the input/output variables, membership functions, inference rules, operators, and methods used.

## 3.3 Research Methodology

We selected a sample of three electives viz. data mining (DM), computer networks and security (CNS), and natural language processing (NLP) for analysis and prediction. We obtained data of PESIT students who had just completed these electives within the last two years. After selection and data cleaning, we got 25–30 student data for each elective. We used SMOTE algorithm to increase the data set to 200–300, since some grades were underrepresented. After creating a "arff" data file, we used WEKA Java library to build a C4.5 DT. This DT is then used to predict the elective grade of a student who has taken the necessary prerequisites.

**Table 1** Mapping of grades with grade points

| Grade | S | A | B | C | D | E | F |
|---|---|---|---|---|---|---|---|
| Grade points | 10 | 9 | 8 | 7 | 6 | 5 | 4 |

**Table 2** Mapping of grade points to grade levels

| Grade Level | Very high | High | Medium | Low | Very low |
|---|---|---|---|---|---|
| Grade points | 10–8 | 9–7 | 8–6 | 7–5 | 6–0 |

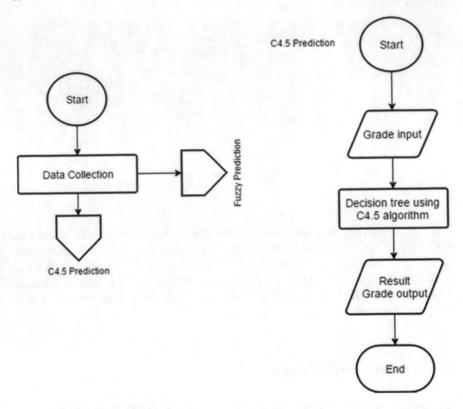

**Fig. 3** Flow chart for C4.5 predictor

The student grades are fuzzified into VERY HIGH, HIGH, MEDIUM, LOW, and VERY LOW. (Refer Tables 1 and 2), and another C4.5 DT is built using these fuzzy attributes. From the DT, rules are extracted (Refer Fig. 1) for each elective performance prediction. An FCL file is created for each elective that contains two fuzzy input variables, one fuzzy output variable, and the rule block containing the inference rules extracted from the DT. We had about a dozen rules in the rule block for each elective. We used the defaults of "min" for "and," "max" for "or," min activation method and max accumulation method. We used "Center of Gravity" defuzzification method.

We used jFuzzyLogic as the fuzzy programming framework. Once the student grades for the prerequisites are entered, the fuzzy logic system (FLS) fuzzifies the grades and executes the rule base. The matched rules are accumulated and COG is calculated by the inference engine. The output variable is then defuzzified and displayed to the student. The elective performance predicted by fuzzy logic is compared to that of the DT. For each elective, we have taken 20% of the data set as test set and compared their performance.

The E-DSS has a front end user interface developed using Spring MVC (model-view-controller) framework. It supports two types of users: the administrator and

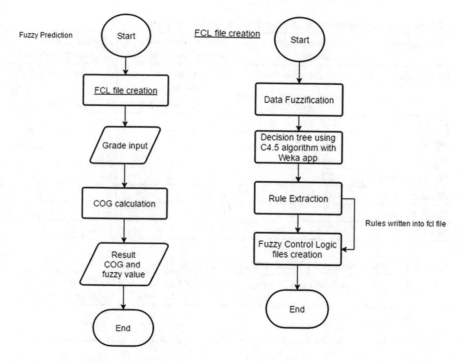

**Fig. 4** Flow chart for FLS predictor

the student user. The administrator can load the data set for the electives and the prerequisite courses. The student user can enter his past course performance (grades) and specify the electives of interest. The C4.5 DT predictor will predict the grade of the interested elective if the prerequisites are satisfied. The FLS will predict both the grade level and the grade so the student is aware of his potential. The logical flow of the overall system, C4.5 predictor, and the FLS predictor is shown in Figs. 3 and 4, respectively.

## 4 Results and Discussion

The results of C4.5 decision tree constructed and tested based on PESIT student data are shown in Table 3. Elective grade prediction accuracy was 85% for DM, 79% in CNS, and 68% in NLP. It was found that a good number of predictions were made to the adjoining classes (e.g., In CNS, 14 students securing A grades were predicted with S). The lower numbers for NLP are mainly because the data set was imbalanced, and there were not many student data for lower grades. We also tested the fuzzy logic system using 20% of the data set, and the results are summarized in Table 4. We found that almost 50% of the errors were due to fuzzy logic prediction in

**Table 3** Elective performance prediction results using decision tree

| Decision tree | Elective (data mining) | Elective (computer network and security) | Elective (natural language processing) |
|---|---|---|---|
| Prerequisites | 1. Data structures | 1. Computer networks | 1. Finite automata |
| | 2. Algorithms | 2. Operating systems | 2. Algorithms |
| Data set | 216 | 193 | 203 |
| Accuracy (%) | 85 | 79 | 68 |
| Kappa | 0.82 | 0.75 | 0.63 |
| Wt. avg precision | 0.9 | 0.8 | 0.6 |
| Wt. avg recall | 0.847 | 0.788 | 0.68 |
| Wt. F-measure | 0.826 | 0.756 | 0.6 |
| Wt. ROC area | 0.954 | 0.945 | 0.92 |

**Table 4** Elective performance prediction results using fuzzy logic

| Fuzzy logic | Elective (data mining) | Elective (computer networks and security) | Elective (natural language processing) |
|---|---|---|---|
| #Test cases | 30 | 23 | 28 |
| Accuracy (%) | 73 | 65 | 57 |

the adjoining grade levels. Choosing three grade levels instead of five would improve prediction.

Overall, the results are encouraging for the performance prediction for electives. Not surprisingly, students who performed successfully well at the required prerequisite courses have also performed well in the related elective courses. However, there are several cases where students who performed average in one of the prerequisites, end up with good elective performance prediction. Such information is encouraging for students to plan and decide.

We have assumed that the performance in the elective course is completely dependent on the performance of related core courses (prerequisites). In reality, there can be other factors influencing the elective, including the time gap between the courses, instructor/syllabus and method of valuation. We have ignored these factors while modeling both the DT and the FLS. Rules based on these additional factors can be added to the FLS to improve the prediction.

# 5 Conclusion

Choosing an elective course can be daunting for a student since there are several factors to be considered. A decision support system that can predict the student performance in the elective course, based on historical data of other students, can be very helpful. We have successfully implemented a decision support system using a decision tree and fuzzy logic. The E-DSS can predict the range of performance and also tentative grade in the elective. Data from PESIT computer science students were used to train and test the DSS. The results indicate that it is possible for student to get a fair idea of their performance in the elective even before choosing it. This information can be combined with other factors before actually choosing an elective.

**Acknowledgements**  We would like to thank PESIT, Bangalore, for providing us the student data set to conduct this experiment. The authors would also like to thank Prof. Natarajan for his valuable suggestions and encouragement.

# References

1. Bendakir, N., Aimeur, E.: Using association rules for course recommendation, American Association for Artificial Intelligence (2006)
2. Golumbic, M.C., Markovich, M., Tsur, S., Schild, U.J.: A knowledge-based expert system for student advising. IEEE Trans. Edu. **E-29**(2), 120–124 (1986)
3. Kristiansen, S., Sorensen, M., Stidsen, T.R.: Elective course planning. Eur. J. Oper. Res. **215**, 713–720 (2011)
4. Parameswaran, A., Venetis P., Molina H.G.: Recommendation systems with complex constraints: a course recommendation perspective. ACM Trans. Inf. Syst. (2011)
5. Farzan, R., Brusilovsky, P.: Social navigation support in a course recommendation system. In: Adaptive Hypermedia and Adaptive Web-Based Systems, pp. 91–100. Springer, Berlin Heidelberg (2006)
6. Klasnja-Milicevic, A., Vesin, B., Ivanovic, M., Budimac, Z.: E-Learning personalization based on hybrid recommendation strategy and learning style identification. Comput. Edu. **56**, 885–899 (2011)
7. Aher, S.B., Lobo, L.M.R.J.: Combination of machine learning algorithms for recommendation of courses in e-learning system based on historical data. Knowl. Based Syst. **51**, 1–14 (2013)
8. Sobecki, J., Tomczak, J.M.: Student courses recommendation using ant colony optimization. Intell. Inf. Database Syst. **5991**, 124–133 (2010)
9. Harsiti, M.A., Sigit, H.T.: Implementation of fuzzy-C4.5 classification as a decision support for students choice of major specialization (IJERT). Int. J. Eng. Res. Technol. **2**(110) (2013)
10. Adak, M.F., Yumusak, N., Taskin, H.: An elective course suggestion system developed in computer engineering department using fuzzy logic. In: Industrial Informatics and Computer Systems International Conference on (CIICS), pp. 1–5 (2016)
11. Quinlan, J.R.: Induction of decision trees. Mach. Learn. **1**(1), 81–106 (1986)
12. Quinlan, J.R.: C4.5: programs for machine learning. Elsevier, Armsterdam (2014)
13. Hssina, B., et al.: A comparative study of decision tree ID3 and C4.5. Int. J. Adv. Comput. Sci. Appl. **4**(2) (2014)
14. Eibe, F., Hall, M.A., Witten, I.H.: The WEKA Workbench. Online Appendix for "Data Mining: Practical Machine Learning Tools and Techniques", 4th edn. Morgan Kaufmann, Burlington (2016)
15. Mendel, J.M.: Fuzzy logic systems for engineering: a tutorial. Proc. IEEE **83**(3), 345–377 (1995)
16. Cingolani, P., Alcalá-Fdez, J.: jFuzzyLogic: a java library to design fuzzy logic controllers according to the standard for fuzzy control programming. Int. J. Comput. Intell. Syst. **6**(sup1), 61–75 (2013)

# An Appearance-Based Gender Classification Using Radon Features

Ratinder Kaur Sangha and Preeti Rai

## 1 Introduction

Face is a characteristic feature of the human beings which contains identity, age, and emotions. Gender classification from a person's face could play an important role in computer vision such as security surveillance systems, search engine, demographic studies, marketing research and performance enhancement (face recognition, smart human–computer interface). In real-world scenario, due to natural reasons, images might be occluded naturally like injury, wearing scarf, or sunglasses because of weather conditions [1] and thus, it becomes difficult to classify gender from such a partial occluded face. In this work, wavelet and Radon transforms are combined together for extracting features to classify male or female from facial information.

The basic gender classification system used for our work is shown in Fig. 1 contains mainly three modules, i.e., preprocessing, features extraction, and classifier. In a pre-processing, basically relevant features are taken out, which are the most potential segment of an image. Feature extraction is done by combing wavelet and Radon transforms. The obtained features are then passed to a powerful supervised learning algorithm using SVM to discriminate male and female.

This paper is assembled as follows: A review of the past research work in the area of gender classification is given in Sects. 2 and 3 describes our gender classification technique using wavelet and Radon transforms as features extraction and SVM as classifier. Experimental analysis and conclusion are shown in Sects. 4 and 5, respectively.

R. K. Sangha (✉) · P. Rai
Gyan Ganga Institute of Technology and Science, Jabalpur, India
e-mail: ratinder18sangha@gmail.com

© Springer Nature Singapore Pte Ltd. 2019
R. K. Shukla et al. (eds.), *Data, Engineering and Applications*,
https://doi.org/10.1007/978-981-13-6347-4_15

**Fig. 1** Gender classification
system

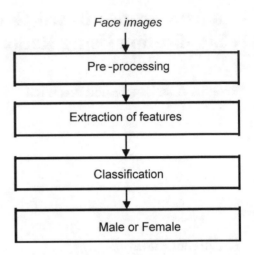

Face images

Pre-processing

Extraction of features

Classification

Male or Female

## 2 Review of Literature

This chapter presents a review of available and relevant literature. In gender classification, the two main approaches to detect the features of the face are appearance based [2] and geometrical based [3]. The chapter investigates various techniques developed under these methods and finally summarizes the literature review, with the identification of the shortcomings and opportunities provided by the past research. In the prior works [4–7], raw face image had been used for classification, and high classification rate was obtained. Later, new techniques were developed for the extraction of features and classification of gender. Sun et al. [8] used local binary patterns (LBP) histograms along with Adaboost for classification under constrained environment to obtain 95.75% classification rate. Makinen and Raisamo [9] used alignment and non-alignment approaches to compare different gender classification schemes. Caifeng [10] used discriminative LBP-Histogram (LBPH) on LFW to achieve 94.81% classification rate. Berbar [11] used three methods for feature extraction, i.e., discrete cosine transform (DCT), gray-level co-occurrence matrix (GLCM), and wavelet transform to extract features and SVM as classifier. It provides better classification accuracy for AT@TAT@T, Faces94, UMIST, and less accurate for FERET.

Rai and Khanna [12] recommend a system based present on $(2D)^2$PCA on real Gabor space by which they achieve better classification rate over 2DPCA either in horizontal or vertical direction, for smaller features size. Rai and Khanna [2] proposed a system using Radon and wavelet transforms with higher classification rate. In this paper, they provide comparative study of a face image by Radon and wavelet transforms with DCT for small features vectors and KNN as classifier. Juili et al. [13] used a novel local texture pattern based on zigzag scanning pattern, to extract the texture from the images. This experiment was done on FERET database and provides improvement in the recognition performance. It also provides a comparison study with other methods. Recently, Selvaraj [14] applied a basic algorithm PCA, linear discriminate analysis, and SVM for gender classification and then discusses their comparative study of the methods.

Xu et al. [15] recognized the gender classification using multi-scale local binary pattern histogram. They extracted features by multi-scale LBPH, which represents richer local and global interested information of facial images in DoG (Difference of Gaussian) space with SVM as classifier, accuracy rate of 97.7% for ICI dataset and 96.4% for FERET(1490 males, 920 female images). Zheng et al. [16] recommend a system for classification of age and gender from face images by local directional pattern (LDP) in three steps, preprocessing, extraction features using LDP algorithm, store LDP histogram, and classification using SVM. They have applied LDP and achieved 95% recognition rate for gender dataset. Rai and Khanna [17] combined approximation face image (AFI) with PCA and $(2D)^2$PCA for features extraction and SVM as classifier. In the proposed system, AFI+ $(2D)^2$PCA achieved better performance than AFI+PCA for small feature size, under conditional as well as unconditional environments.

Lale and Karande [18] implemented PCA technique with minimum distance classifier for classification of gender on FEI face database of 100 people (50 male and 50 female) and obtain 97% accuracy. Anusha et al. [3] classified gender classification along with expression (anger and joy) with salient facial patches and also working well on low-resolution images. Nineteen patches are extracted from the face and based on facial patches, gender is detected. In this method, eye and lips are detected by Haar classifier and Sobel Edge detector, gender is predicted by QDA on JAFFE and CK+ database. Goel and Vishwakarma [19] compared the proposed techniques with the existing technique [11]. Berbar used DWT for feature extraction, which gives less accuracy as compared to the proposed technique. In the proposed work, they used KPCA followed by SVM-RBF and performance obtained is 97.35% on AT@T database, 96.67% on Georgia Tech face database, and 99.7% on Faces94 database.

Rai and Khanna [20] proposed a system working for both occluded and normal face images with Gabor filter, two-dimensional 2DPCA, and SVM as classifier.

From this approach, it is concluded that performance from the system is improved by targeting Approximation Face Sub-Images as compared to full face images, 90% accuracies system renders for lower occlusion conditions and 86.8% accuracy in higher occlusion conditions.

# 3   Proposed Work

The proposed work shown in Fig. 2 consists of three modules, i.e., preprocessing, feature extraction by Radon and wavelet transforms, and classification by SVM.

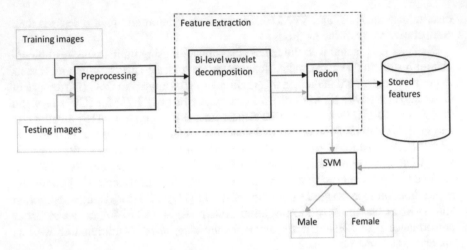

**Fig. 2** Workflow of the proposed system

## 3.1 Preprocessing

In preprocessing module, relevant features are extracted from the face image as it contains more useful information. To obtain such features, manually selecting three points of the face, i.e., first points on left corner and second point on right corners of the eyes, whereas center point of chin is considered as third point [1, 20]. By connecting the entire points together, triangle is obtained as given in Fig. 3.

In the proposed work, '$P$' is marked as centroid and '$r$' as the distance between $P$ and any one of the three points marked earlier to form a triangle; then square of $2r$ side is constructed. Region of the face image obtained inside the square is called desired relevant information and finally color conversion into the grayscale is done. This preprocessing procedure is used in this work for extracting useful features from the original face image which may be normal or occluded.

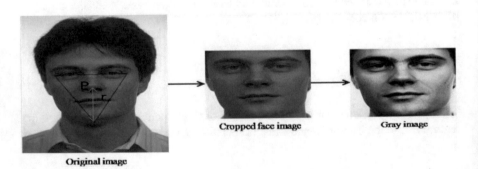

**Fig. 3** Preprocessing (cropped face image from FERET)

## 3.2 Feature Extraction Techniques

In feature extraction module, wavelet and Radon transforms are used together to find the facial features of male and female, and finally, SVM is used for classification.

### 3.2.1 Wavelet Transform

The wavelet decomposition is a kind of time–frequency signal analysis method [21]. By using it, the face image can be decomposed into many sub-band images with different spatial resolution frequency characteristic and directional features [22]. When a separable transform is applied, only the approximation coefficients may need further decomposition. If required, approximation coefficients (LL) can be further decomposed into four sub-bands by using identical filter bank at the second level of decomposition. Wavelet transform has a high time and frequency resolution for stationary as well as non-stationary signals. In this work, the face image is decomposed up to two levels. Not more than two-level decompositions of face are considered as it results in loss of information. Approximation face image (AFI) obtained contains all the global information from the face. Hence, only approximation coefficient (AFI) is taken for further procedure.

### 3.2.2 Radon Transform

Introduction to Radon transform was given by Austrian mathematician Johann Radon in 1917. Inherent properties of Radon transform to make it useful tools for capturing directional features of the image. Further, it is used to compute the projection of an image along specified directions [2] as it is rotation and translation invariant to preserve variations in pixels intensities. Implementing Radon transforms on an image $f(x, y)$ for a particular set of angle can be viewed as evaluating the projection of the image along the given angles. The resulting new image is $R(r; \theta)$. Projection of an image $f(x; y)$ along an angle $\theta$ is as shown in Fig. 4.

Radon transform for an image $f(x; y)$ along angle $\theta$ is defined as

$$R(r;\theta) = \int_{-\infty}^{\infty} \int_{-\infty}^{\infty} f(x;y)\delta(r - \cos\theta - y\sin\theta)dxdy$$

where the $\delta(\cdot)$ Dirac delta function is the perpendicular distance from a line from the origin and $\theta \in [0, \pi]$ is the angle formed by the distance vector. Radon transforms has a wide application in textural classification, computer tomography, and local feature in edge detection. Currently, Radon transforms has been successfully used much for face recognition or medical images.

**Fig. 4** Radon projection of an image

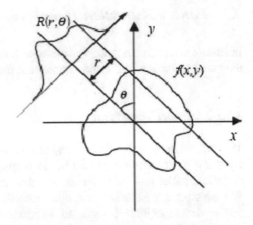

## 3.3 Classification

Gender classification is two-class problems, and literature also supports the use of SVM for gender classification. Among all the classifier, SVM has given the best generalization result and falls within the category of the supervised learning algorithm [23, 24]. SVM is based on the concept of nonlinear mapping the original training data into a higher-dimensional feature space that aims to separate a set of objects with maximum distance.

Basic idea of SVM finds the optimal classification hyperplane that separates those data into different categories. After extracting the features from image, SVM determines some support vector from the feature space. These vectors help to find the optimal hyperplane. SVM classification generates a decision boundary based upon the training set, which helps in predicting the target value from the testing dataset.

SVM is a linear classifier that uses the linear hypothesis to classify the data. If data is linearly non-separable, then it maps the data into higher dimensions using kernel function. Different types of the kernel are linear, polynomial, quadratic, radial basis function, etc. To classify gender of images, RBF kernel is most widely used. As RBF has been less numerical difficult and can easily map non-linear samples in higher dimensional.

## 4 Experimental Analysis

The proposed work is evaluated on Intel Core @ 2.10 GHz Processor with 2 GB memory RAM. All proposed work are performed on MATLAB R2013a. Performance of the system is measured by the classification rates obtained from the system. The experiment is conducted over 784 face images (393 males, 391 females), while

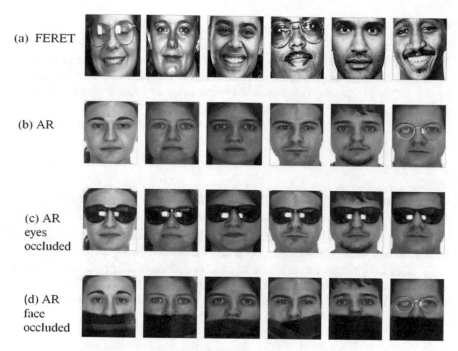

**Fig. 5** Sample of face images

images used are frontal face images and varied extends of occlusion, introduced through wearing sunglasses and scarf.

Databases used for experimental analysis are available publicly with a satisfactory number of male and female images. FERET [25] used consists of all frontal face images with different variation in poses and expression as shown in Fig. 5. Face images for AR [26] databases are selected to form three datasets. AR frontal contains only frontal images, AR expressions contain with light distortions, and AR occluded with higher distortions (contains eyes and faces occluded). Some preprocessing used face images are shown in Fig. 5. Table 1 contains outline details of databases used in the experimental analysis.

**Table 1** Facial databases used for experimental evaluation

| Databases | Total no. of face images | Characteristics |
|---|---|---|
| FERET | 484 | Frontal gray color facial images with variation in Expression |
| AR | 100 | Neutral frontal facial images |
| AR face occluded | 100 | Frontal face images occluded with scarf |
| AR eyes occluded | 100 | Frontal face images occluded with glasses |

## 4.1 Classification Rate for Normal Face Images

The FERET contains total 484 face images, where each image is of size $231 \times 231$ pixels. AR databases used to contain a total of 100 frontal face images corresponding to size $120 \times 165$ pixels. The proposed system achieves 97.68% classification rate for FERET and 97.22% for AR database as shown in Fig. 6.

Different result is obtained in a different database as there is the variation in pose and expression.

## 4.2 Classification Rate for Partially Occluded Face Images

Results for occluded AR (of size $120 \times 165$ pixels) database with occlusion introduced through wearing sunglasses and scarves are shown in Fig. 7.

Total of 200 images are used; these images consist of 100 tops occluded (i.e. eyes occluded) face images (50 females, 50 males) and 100 bottom occluded face images (50 females, 50 males). The classification rate for lower part (wearing scarf)

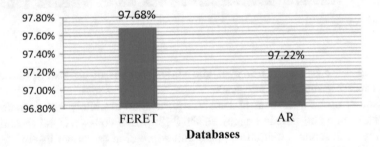

**Fig. 6** Classification rate for normal face images

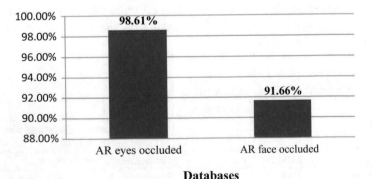

**Fig. 7** Classification rate for partially occluded face images

**Table 2** Classification rate for cross-database

| S. No | Database used in training | Database used in testing | Classification rate (%) |
|-------|---------------------------|--------------------------|-------------------------|
| 1 | AR | AR face occluded | 63.88 |
| 2 | FERET | AR face occluded | 56.94 |
| 3 | AR expression | AR eyes occluded | 95.83 |
| 4 | FERET | AR eyes occluded | 62.50 |
| 5 | AR | AR eyes occluded | 94.44 |

is 91.66 and 98.61% for upper part (wearing glasses). It concluded that in the gender classification system, the lower area of face (i.e., mouth and chin) contains more discriminative features as to the upper area (i.e., eyes).

## 4.3 Classification Rate for Cross-Databases

Correctness and computational time on the system are checked for proposed techniques on cross-databases. In order to evaluate classification rate for cross-database, the proposed system is trained with FERET and AR databases. Training databases consist of non-occluded face images with variation in expression and pose, while testing has been done with AR occluded images taken under unconditional environment. As shown in Table 2, for cross-database system achieved higher classification rate, i.e. 95.83% for AR expressions faces images in training and AR eyes occluded in testing. It has been concluded that setup is successfully achieved more than 56.94% of the classification rate accuracy for cross-database evaluation.

## 5 Conclusion

In this paper, wavelet and Radon transforms are used together, to calculate the robust feature for classifying gender. This system works on gender classification by extracting features from expression varied and partially occlusion faces images. Experimental results show that the proposed system achieved 98.61% accuracy for AR eyes occluded and 91.66% for AR faced occluded. Figure 7 indicates that the lower surface of face contains additional discriminative features regarding eyes. For cross-database, the performance from the system achieved average classification rate of 56% for normal dataset into training and occluded dataset in testing (under conditional and unconditional environment).

The future work is to handle gender classification with video-based images, as the proposed system includes only still images to accelerate this process.

# References

1. Rai, P., Khanna, P.: An efficient and robust gender classification system. In: International Conference on Computational Intelligence and Communication Networks, pp. 254–261. IEEE (2015)
2. Rai, P., Khanna, P.: Gender classification using radon and wavelet transforms. In: IEEE 5th International Conference on Industrial Information System 2010, pp. 448–451. 29 July–01 Aug 2010
3. Anusha, A.V., Jayasree, J.K., Bhaskar, A., Aneesh, R.P.: Facial expression recognition and gender classification using facial patches. In: International Conference on Communication Systems and Networks, pp. 200–204. IEEE, 21–23 July 2016
4. Golomb, B.A., Lawrence, D.T., Sejnowski, T.J.: Sexnet: a neural network identifies sex from human faces. In: Advances in Neural Information Processing Systems, pp. 572–577. Colorado, USA (1991)
5. Wiskott, L., Fellous, J.M., Kruger, N., von der Malsburg, C.: Face recognition and gender determination. In: International Workshop of Automatic Face and Gesture Recognition, pp. 92–97 (1995)
6. Tamura, S., Kawai, H., Mitsumoto, H.: Male/female identification from $8 \times 6$ low resolution face images by neural networks. Pattern Recogn. **29**(2), 331–335 (1996)
7. Moghaddam, B., Yang, M.H.: Learning gender with support face. IEEE Trans. Pattern Anal. Mach. Intell. **24**(5), 707–711 (2002)
8. Sun, N., Zheng, W., Sun, C., Zou, C., Zhao, L.: Gender classification based on boosting local binary pattern. In: Advances in Neural Networks, pp. 194–201. Springer (2006)
9. Makinen, E., Raisamo, R.: An experimental comparison of gender classification methods. Pattern Recogn. Lett. **29**(10), 1544–1556 (2008)
10. Caifeng, S.: Learning local binary patterns for gender classification on real-world face images. Pattern Recog. Lett. **33**, 431–437 (2012)
11. Berbar, M.A.: Three robust features extraction approaches for facial gender classification. Visual Comput. **30**, 20–31 (2014)
12. Rai, P., Khanna, P.: An illumination, expression, and noise invariant gender classifier using two-directional 2DPCA on real Gabor space. J. Visual Lang. Comput. **26**, 15–28 (2015)
13. Juili, Y., Lai, C.C., Wu, C.H., Pan, S.H., Lee, S.J.: Gender classification from face images with local texture pattern. Int. J. Ind. Electron. Electr. Eng. 15–17 (2015)
14. Selvaraj, V.: Gender classification for digital solutions using facial images. In: Third International Conference on Images Information Processing, pp. 502–505. IEEE (2015)
15. Xu, Y., Zhao, Y., Zhang, Y.: Multi-scale local binary pattern histogram for gender classification. In: 8th International Congress on Image and Signal Processing, pp. 654–658. IEEE (2015)
16. Zheng, Y., Hu, M., Ren, F., Jiang, H.: Age Estimation and Gender Classification of Facial Images based on Local Directional Pattern, pp. 103–107. IEEE (2014)
17. Rai,P., Khanna, P.: Appearance Based Gender Classification with PCA and $(2D)^2$ PCA on Approximation Face Images. IEEE (2015)
18. Lale, D.P., Karand, K.J.: Gender Classification Using Facial Features. IJARECE, pp. 2227–2231 (9 Sept 2016)
19. Goel, A., Vishwakarma, V.P.: Gender classification using KPCA and SVM. In: International Conference on Recent Trends in Electronics Information Communication Technology, pp. 291–295. IEEE, 20–21 May 2016
20. Rai, P., Khanna, P.: A gender classification system robust to occlusion using Gabor features based $(2D)^2$ PCA. J. Visual Commun. Image Represent **25**(5), 1118–1129 (2014)
21. Kalsi, K.J, Rai, P.: Gender classification of emotion and gender using approximation image Gabor local binary pattern. In: 2017th International Conference on Cloud Computing, Data Science and Engineering, pp. 623–628. IEEE (2017)
22. Daubechies, I. et al.: Ten lectures on wavelets. SIAM **61** (1992)
23. Howley, T., Madden, M.G.: The genetic kernel support vector machine description and evaluation. Artif. Intell. Rev. **24**(3–4), 379–395 (2005)

24. Vapnik, V.N.: Statistical Learning Theory. Wiley, Hoboken (1998)
25. Phillips, P.J., Wechsler, H., Huang, J., Rauss, P.J.: The FERET database and evaluation proce-
    dure for face-recognition algorithms. Images Vision Comput. **16**(5), 295–306 (1998)
26. Martinez A.M.: The AR Face Database, CVC Technical Report 24
27. Gudla, B., Chalamala, S.R., Jami, S.K.: Local binary patterns for gender classification. In:
    3rd International Conference on Artificial Intelligence, Modelling and Simulation, pp. 19–22.
    IEEE (2015)

# An Automatic Text Summarization on Naive Bayes Classifier Using Latent Semantic Analysis

Chintan Shah and Anjali Jivani

## 1 Introduction

Automatic text summarization is a process to reduce the text in a system and to generate a good summary [1]. Natural language processing and machine learning are the major areas for text summarization. This paper presents the idea of text summarization on basis of text extraction.

Automatic text summarization is the process of creating a shorter version of a given source such as news article, which retains the main information of source, so user can understand the concept of article. There are two types of summarization, namely abstractive and extractive summarization. In extractive summarization, the process involves using specific techniques to select sentences which have the highest score in the retrieved document and consolidating all the extractive sentences on the basis of score and building the summary, whereas in the abstractive summary, the original text is converted into a similar semantic form with the help of linguistic methods to get a comprehensive version of the original document. Latent semantic analysis is used for the dimension reduction, and once the summary is ready, Naïve Bayes classifier is used for training the model. The summary is predicted once the model is ready. The primary step of text summarization is to identify the important features. After that, preprocessing that includes sentence segmentation, tokenization, stop word removal, and stemming plays a major role. After performing preprocessing step and once the corpus is ready, the document matrix and tf-idf matrix are built on corpus for each term. Singular-value decomposition (SVD) is applied to tf-idf matrix and different types of concept are ready. In concept, on basis of threshold, value given

C. Shah (✉)
Shankersinh Vaghela Bapu Institute of Technology, Gandhinagar, India
e-mail: chintan.shah84@gmail.com

A. Jivani
The Maharaja Sayajirao University of Baroda, Vadodara, India
e-mail: anjali_jivani@yahoo.com

© Springer Nature Singapore Pte Ltd. 2019
R. K. Shukla et al. (eds.), *Data, Engineering and Applications*,
https://doi.org/10.1007/978-981-13-6347-4_16

for prediction is either 1 (part of summary) or 0 (not part of the summary). There are many concepts present, but since not all are useful, recursive feature elimination techniques to select only the important concepts are used in the present study. Naïve Bayes theorem is applied on the selected concepts for training and predicting the summary, which is generated on the basis SVD calculation.

Section 2 of the paper describes the related work done on automatic text summarization. Section 3 shows the proposed method for the improvement of text summarization, and Sect. 4 presents the results and discussion. Conclusions are covered in Sect. 5.

## 2   Related Work

Lunh (1958) [2] introduced first text summarization based on term frequency. In 1969, automatic text summarization had used some standard methods to assign sentence weights. Such methods are cue, title, and location method. In the early 1990s, the machine learning techniques were applied on natural language processing (NLP) with the help of statistical methods to generate extractive summary on given corpus [3]. Hidden Markov model, linear regression, Naïve Bayes classification, and clustering techniques were used for the improvement of extractive summary. Artificial intelligence has taken importance in automatic text summarization.

In latent semantic analysis (LSA), many researchers have applied a variety of concepts on SVD to achieve comprehensive summary. Gong and Lui (2001) proposed method for text summarization on Latent Semantic Analysis (LSA), after calculating SVD value on input matrix of document, the $V^T$ matrix, extracted concepts and X sentences is used for choosing important sentences. Steinberger and Jezek (2004), proposed method is extension of Gong and Lui. They used both $V$ and $\sum$ matrices for sentences selection. Murray et al. (2005) suggested method, which two steps of LSA algorithms implemented before selection step. Multiple important concepts are selected for the sentence selection.

## 3   The Proposed Method

The aim of automatic text summarization is to generate important sentences for summaries. The proposed method uses statistical methods, i.e., singular-value decomposition (SVD), probability, feature ranking with recursive feature elimination on generated concepts on SVD, Naïve Bayes machine learning algorithm for training documents, and prediction. The flow of the proposed method is shown below (Fig. 1).

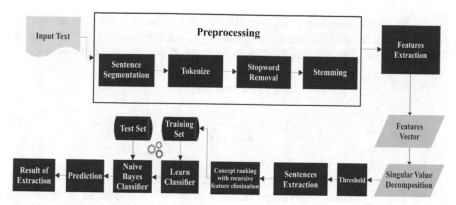

**Fig. 1** Proposed model

Input: An input document.
Output: A summarized text as per compression ratio.

i.  Read input text file.
ii. Preprocess the file.

    a. Remove all unnecessary characters.
       In this step, all unnecessary characters like punctuations and symbols will be removed.
    b. Convert all word into lower case.
       All words are converted into lower case with Python built-in function lower().
    c. Split each word by sentence—segmentation.
       Segmentation is the task where text is divided into word, unit, or topic.
    d. Tokenize each word using Porter Stemmer.
    e. Remove all stop words.

iii. Feature extraction.

    a. Occurrence of a word in a file.
       It is known as term-document matrix. A mathematical matrix explains the occurrence of term in a collection of text. Word (or $n$-gram) frequencies are typical units of analysis when working with text collections. It is term-document matrix and a vocabulary list. It converts a collection of text document to matrix of token counts. At process, if system does not provide a priori dictionary and analyzer, then the system can use feature selection as equal to vocabulary size of analyzing data. When preparing a matrix, rows represent the document and columns represent terms.
    b. From occurrence to frequencies (tf-idf).
       Term frequency–inverse document frequency is a numerical method to understand importance a word is in corpus. The tf-idf value increases regularly to the number of times a term appears in the document, but is often compensated

for by the frequency of the word in the corpus, which helps to adjust the fact that some words appear more frequently in general [wiki]. Different types of tf-idf weighting methods are used for scoring and ranking a document.

$$S = TF * IDF \tag{1}$$

$$TF_i = \frac{T_i}{\sum_{k=1}^n T_k}, IDF = \log \frac{N}{n_i} \tag{2}$$

iv.  Latent semantic analysis (LSA).

Latent semantic analysis (LSA) is one of the statistical techniques for extracting the meaning of contextual usage of words by statistical computations applied to a large corpus of text. The principal aim is that the information about all the word contexts in which a given word appears provides a set of mutual constraints that largely determines the similarity of meaning of words and set of words to each other. The adequacy of LSA's reflection of human knowledge has been proven in a variety of ways.

Latent semantic analysis (LSA) is a statistical model that compares semantic similarity between fragments of textual information for word usage. It used for improving the efficiency for methods of information retrieval. By using LSA, the problem of synonymy, in which a different word or term can be used to explain the same semantic concept, can be solved. LSA is also used to analyze the relationships between the pair documents and their terms, which it contains by producing a set of concepts related to documents and terms. LSA accepts that words, which are close in meaning, will occur in similar pieces of text. A matrix includes rows and columns, rows will represent unique terms from document, and columns will represent each paragraph. A matrix is built from text. Moreover, we can truncate rows with the help from singular-value decomposition (SVD), which is a mathematical technique, which conserves the resemblance structure among columns.

LSA has three main steps, which are described below.

1.  Creation of input matrix
2.  SVD—Singular-value decomposition
3.  Selection of sentences.

**Input Matrix Creation**

The input needs to be presented in such a way that computer can understand and do calculation as necessary. For that, representation can be done via matrix, where columns are represented as documents/paragraphs and rows are represented as unique words/terms, which appear in documents. In matrix, a cell indicates the importance of the word in sentence. Various approaches can be used for filling the cell values, but words do not appear in all documents, and hence, the so-created matrix will become sparse matrix [4].

For summarization, the input matrix is significant because it directly effects on the result of SVD. As SVD is very complex technique, the complexity increases with the size of input matrix. To reduce the matrix, the words can be reduced by various ways such as removing stop words, punctuation marks, and tokenization. Various approaches can be used for filling cell values, which are described below.

a. Frequency of word: The frequency of word in sentence value filled in cell values.
b. Binary representation: the value of cell is fill with either 0 or 1 on the being of a word in sentence.
c. Tf-idf—term frequency–inverse document frequency. With this method, we can fill cell values. Higher values show that words that are more common appear in sentence but not in the documents. A higher value also indicates that word is more representative for particular sentence.
d. Log entropy: The cell value is filled with the amount of information that can be held by the sentence.

**Singular-Value Decomposition**

SVD is a statistical model that shows relationship among words/terms and sentences. It decomposes the input matrix into three other matrices as shown below

$$A = U\Sigma V^{\mathrm{T}}$$

$A$   Input matrix ($m \times n$)
$U$   Words $\times$ extracted concepts ($n \times n$)
$\Sigma$   Scaling values, diagonal descending matrix ($n \times n$)
$V^{\mathrm{T}}$   Sentences $\times$ extracted concepts ($n \times n$) (Fig. 2).

**Sentences selection**

To select relevant sentences using the singular-value decomposition results, various approaches and algorithms such as Gong and Liu approach, Steinberger and Jezek's approach, Murray, Renals, and Carletta's approach are proposed. In the

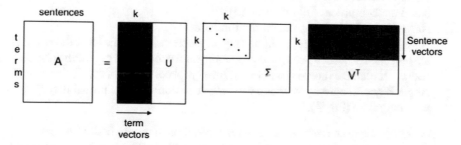

**Fig. 2** Singular-value decomposition diagram [3]

**Fig. 3** Algorithm for
threshold selection

```
For each generate concept
        Begin
                Check whether concept hold specific
        threshold
                If yes
                        Set the field as 1 for summarization
                Else
                        Set the filed as 0
        End
End for
```

present study, Gong and Liu approach for summarizing the paragraph that uses $V^T$ matrix for sentence selection is adopted.

v.  Generation of Summary.

After performing SVD, $V^T$ matrix, the matrix of extracted sentences X concepts is used for picking the significant sentences. In $V^T$ matrix, row represents the importance of concepts. The cell values demonstrate the relationship between the sentence and the concept. A higher value indicates that concept is more relevant to concept.

The sentences are marked with 1 or 0 with a specific threshold for the prediction of extracted sentences (Fig. 3).

vi.  Select important concepts on basis of Feature ranking with recursive feature elimination.

After performing SVD, it populates many concepts on each sentence. Many times, it is not necessary to select all concepts for prediction of summary. Since each concept does not have a significant importance, it is necessary to select the important concepts only.

Here, we use recursive feature elimination; it is a comprehensive optimization algorithm, which aims to find the best subset on the basis of the performance. It iterates through entire list and prepares a model. It keeps aside best- and worst-performing feature at each repetition. Then, it builds next model with rest concept/features until all concepts are exhausted. Then, it ranks the concepts based on their order of elimination [5].

vii.  Naïve Bayes for training and prediction model.

Naïve Bayes algorithm is based on Bayes' theorem. It assumes that the features are independent. It acquires prior probability and conditional probability on each feature. It predicts the class label by highest probability.

As per Bayes' theorem provides a way of calculation posterior probability $P(c|x)$, $p(x)$ and $p(x|c)$ (Fig. 4).

As per the above step of recursive feature selection, dataset divides into sets, i.e., training set and test. Then Gaussian Naïve Bayes is applied on training set and train model for prediction. Using train model, we predict on test data for summarization against generate on basis of Singular Value Decomposition (SVD).

**Fig. 4** Naïve Bayes

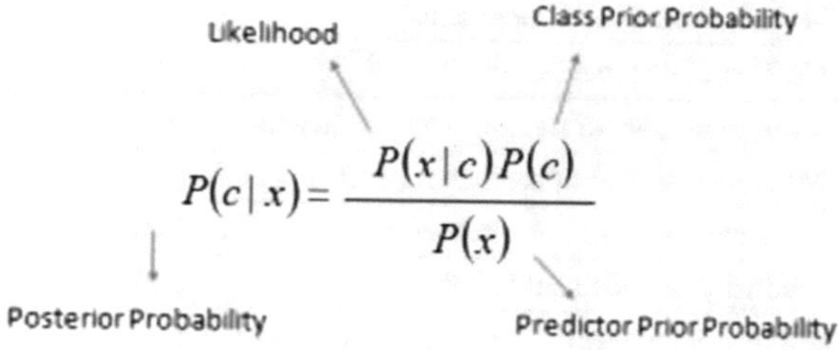

$$P(c \mid X) = P(x_1 \mid c) \times P(x_2 \mid c) \times \cdots \times P(x_n \mid c) \times P(c)$$

## 4  Evaluation Result

Mainly, there are three types of criteria that can be used for the evaluation of summaries: (1) co-selection, (2) content-based similarity, and (3) relevance correlation. Co-selection includes precision, recall, and $F$-measure [6]. A co-selection works only on extractive summary. Content-based similarity will check similarity measure in document. It uses word overlap, longer common subsequence, and cosine similarity [7].

ROUGE 2.0 evaluation toolkit works on criteria of intrinsic summarization. It is used to calculate the ratio of how the reference summary overlaps the system summary. ROUGE evaluation measures generate three types of value for each summary: average precision, average recall, and average $F$-measure.

Precision (Rijsbergen 1979) defines how many retrieved selected sentences are relevant to user's information.

$$precision = \frac{|\{relevant\ sentences\} \cap \{retrieved\ sentences\}|}{|\{retrieved\ sentences\}|}$$

Recall (Rijsbergen 1979) defines how many relevant sentences are selected and successfully retrieved.

$$recall = \frac{|\{relevant\ sentences\} \cap \{retrieved\ sentences\}|}{|\{relevant\ sentences\}|}$$

$F$-measure is considered as harmonic mean (Uddin and Khan 2007) of precision and recall. $F$-measure or $F$-score is defined as.

$$F = \frac{(2 \cdot precision \cdot recall)}{(precision + recall)}$$

The text corpus used in this project includes ten articles from different sources, such as yoga, sports article, movie review, and story. The statics of the documents is given below in Table 1.

**Table 1**  Statistics of documents

| Number of documents | 10 |
| --- | --- |
| Average number of sentences per document | 31.5 |
| Minimum number of sentences per document | 21 |
| Maximum number of sentences per document | 56 |
| Summary of document length (%) | 50 |
| Maximum number of sentences per summary | 28 |
| Minimum number of sentences per summary | 8 |

**Table 2**  Scores of each document

| Document | Precision | Recall | F-score |
| --- | --- | --- | --- |
| Document 1 | 0.83 | 0.84 | 0.80 |
| Document 2 | 0.69 | 0.83 | 0.76 |
| Document 3 | 0.94 | 0.92 | 0.92 |
| Document 4 | 0.95 | 0.91 | 0.86 |
| Document 5 | 0.87 | 0.80 | 0.80 |
| Document 6 | 0.69 | 0.83 | 0.76 |
| Document 7 | 0.80 | 0.67 | 0.62 |
| Document 8 | 0.94 | 0.90 | 0.88 |
| Document 9 | 0.85 | 0.89 | 0.82 |
| Document 10 | 0.88 | 0.86 | 0.79 |

Above chart explains the comparison of different algorithms of text summarization. Different algorithms are used and tested on specific documents for evaluation. As per chart, we can see that text rank is showing better performance in all parameter such as precision, recall, and F-score. As per figure, it is clearly seen that text rank is showing better performance in all cases and next Edmundson is showing great variation for evaluation. We have tested our model against existing techniques and evaluated with different evaluation criteria, which is given in Table 2. The details about the content of the documents are shown in Table 1, and the formulas for calculating the precision, recall, and F-score are explained in detail before Table 1 is displayed.

As per above figure and comparison with existing algorithms as per Fig. 5, it is clearly seen that in most document, the results show a better performance in comparison with the existing algorithms. Document 3, 4, and 8 achieved more than 80% precision, recall, and F-score, whereas in few cases, the performance of model shows less performance in document 2 and 7 (Fig. 6).

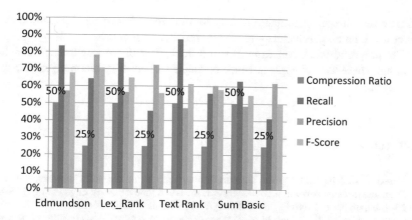

**Fig. 5** Comparison of different summarization tools

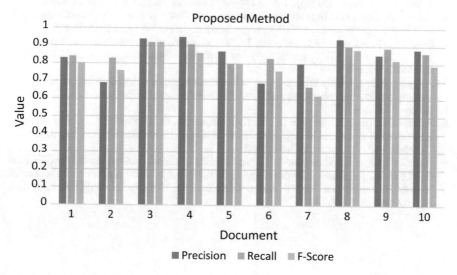

**Fig. 6** Proposed method of evaluation

## 5 Conclusion and Future Work

In this paper, an automatic text summarization approach has been proposed that uses Naïve Bayes classifier on latent semantic analysis. The proposed model works on different type of domains such as international news, sports, review, and story. Another feature of the model is that it uses threshold to select the concepts for summarization. User can choose different values of threshold. With the help of recursive feature elimination technique, only significant concepts are selected for training the data on Naïve Bayes classifier. The test data will be evaluated and predicted for summary after training. To check efficiency with existence algorithms, we have checked on

ROUGE toolkit against human versus system summary. The proposed method shows better result on all parameters—precision, recall, and $F$-score.

The proposed work can extend in the field of artificial intelligence and deep learning, where we can apply advance concepts and generate summary that is more exhaustive.

# References

1. Meena, Y.K., Jain, A., Gopalani, D.: Survey on graph and cluster based approaches in multi-document text summarization. In: Recent Advances and Innovations in Engineering (ICRAIE), 2014, Jaipur, pp. 1–5 (2014). https://doi.org/10.1109/icraie.2014.6909126
2. Gholamrezazadeh, S., Salehi, M.A., Gholamzadeh, B.: A comprehensive survey on text summarization systems. In: 2009 2nd International Conference on Computer Science and its Applications, Jeju, Korea (South), pp. 1–6 (2009). https://doi.org/10.1109/csa.2009.5404226
3. Babar, S.A., Patil, P.D.: Improving performance of text summarization. Procedia Comput. Sci. **46**, 354–363 (2015). ISSN 1877-0509, http://dx.doi.org/10.1016/j.procs.2015.02.031
4. Ozsoy, M.G., Alpaslan, F.N., Cicekli, I.: Text summarization using latent semantic analysis. J. Inf. Sci. (2011). [online] Available at: http://journals.sagepub.com/doi/10.1177/0165551511408848
5. Analytics Vidhya: 6 Easy Steps to Learn Naive Bayes Algorithm (with code in Python) (2017). [online] Available at: https://www.analyticsvidhya.com/blog/2015/09/naive-bayes-explained/. Accessed 23 June 2017
6. Radev, D.R., Teufel, S., Saggion, H., Lam, W., Blitzer, J., Qi, H., Çelebi, A., Liu, D., Drabek, E.: Evaluation challenges in large-scale multi-document summarization: the mead project. In Proceedings of ACL, Sapporo, Japan (2003)
7. Ali, M., Ghosh, M.K., Abdullah-Al-Mamun: Multi-document text summarization: SimWithFirst based features and sentence co-selection based evaluation. In: 2009 International Conference on Future Computer and Communication, Kuala Lumpur, 2009, pp. 93–96 (2009). https://doi.org/10.1109/icfcc.2009.42

# Preserving Patient Records with Biometrics Identification in e-Health Systems

Ambrose A. Azeta, Nicholas A. Omoregbe, Sanjay Misra,
Da-Omiete A. Iboroma, E. O. Igbekele, Deborah O. Fatinikun,
Ebuka Ekpunobi and Victor I. Azeta

## 1 Introduction

Enormous attention has been received in recent years as a result of the deployment of computer technology for health care management system [1]. In other words, the role played by information and communication technology (ICT) in providing health care support services cannot be over-emphasized [2], although with some issues to be addressed. The common issues include—use of easily guessed passwords which has the ability to allow unauthorized access, and also privacy of patient data. The transition of privacy of data (in health data) from the last few decades up until now is such that most computer networks are not only vulnerable but has been compromised

A. A. Azeta (✉) · N. A. Omoregbe · S. Misra · D.-O. A. Iboroma · E. O. Igbekele ·
D. O. Fatinikun · E. Ekpunobi
Center of ICT/ICE Research, CUCRID Building, Covenant University, Ota, Nigeria
e-mail: ambrose.azeta@covenantuniversity.edu.ng

N. A. Omoregbe
e-mail: nicholas.omoregbe@covenantuniversity.edu.ng

S. Misra
e-mail: sanjay.misra@covenantuniversity.edu.ng

D.-O. A. Iboroma
e-mail: damelinks@gmail.com

E. O. Igbekele
e-mail: igbekele.emmanuel@gmail.com

D. O. Fatinikun
e-mail: fatoluwafisayo@gmail.com

E. Ekpunobi
e-mail: eekpunobit9@gmail.com

V. I. Azeta
National Productivity Center, Calabar, Nigeria
e-mail: victaazeta@gmail.com

© Springer Nature Singapore Pte Ltd. 2019
R. K. Shukla et al. (eds.), *Data, Engineering and Applications*,
https://doi.org/10.1007/978-981-13-6347-4_17

in terms of quality of services. When it comes to evaluating the performance of medical information system, security is very important [3]. More so, the ability to monitor adherence from a distant location of patient's health is essential [4]. In addition, eventuality may happen when a misunderstood or incomplete record arise which is traceable to either wrong prescription or medication that could cause patient death. One of the functions of health information system is to minimize the massive manual paper work in medical centers and hospitals and to resolve the reoccurring dearth of health care professionals [3]. Health information system (HIS) could also be prone to medical fraud, missing records, duplicated medical records, strenuous patient authentication and verification system, and inappropriate billing system.

The increasing demand for more reliable authentication system has been reinforced in the health care industry as a result of the push for electronic (*e*) medical records. Almost on a regular basis, several patients in Nigeria go to some clinics to see their doctors. Feedbacks from most of the visits show that such visits mostly duplicate medical records or build on existing once, and therefore introduce redundancy into the medical record system. The current approach of keeping and reorganizing records demands the implementation of reliable system with proficient users' authentication [5]. Patient desires to have the confidence that the privacy of their medical records are in safe hands [6]. Arising from the above, a health information system (HIS) with fingerprint biometrics technology has the potentials to provide some level of data security for patients and health care workers, as has been shown in [7].

A major breakthrough of biometrics solutions in health care is that the technology has been able to transform the hospitals and HIS security system to an appreciable extent. Biometrics has served to be an electronic means of verification and identification of a person based on measurable attributes. The traditional methods often engaged in biometrics solutions include facial recognition, retinal and iris scans, ear shape, fingerprints, voice verification, signature dynamics, hand geometry, skin patterns, body odor, and a lot more. From literature, the use of biometrics is more efficient with specialized devices including cameras that are infrared and used for image acquisition [8]. The low cost of fingerprint biometrics technology has been one of the reasons why most pronounced biometrics system makes use of authentication. Biometrics, powered by technological advancement, (especially the Internet) has resulted in a widespread array of a secured identification, authentication, and verification.

The objective of this implementation-based study is to provide a health information system (HIS) with fingerprint biometrics and password/pin for authentication. The remaining part of the paper is described as follows: Sect. 2 explained in details the literature review while Sect. 3 contains system modeling and design, as well as the user's interface of the developed system. Section 4 presents the discussion and recommendations. In Sect. 5, the final remarks of the paper as well as further works are provided.

## 2   Literature Review

From previous studies, several health information systems (HIS) make use of login password to authenticate patient in the health records management system. However, lots of studies on methods and strategies for fingerprint biometrics are also needed, predominantly those that are able to bring together single-mode factor authentication system. Some of the biometrics-based HISs are discussed in this section.

Lamport developed an authentication scheme using password for authenticating user access remotely [9]. Darrell explained in [10] how the implementation of biometrics is used in identifying security issues and increases information security for doctors, nurses, and patients in health care industries. It also explores the contribution of biometrics in improving security measures within the health care industry for protection of all the entities involved (doctors, nurses, and patient). The research highlighted in [11] developed a smart card system with password authentication scheme using biometrics technique and hash function. This system gave a secured and efficient Telecare system.

The study in [12] gave a comprehensive overview of e-health security challenges using biometrics technology applications for security and privacy issues in the health sector. The study showed a great opportunity for applying biometrics technology in e-health for reliable security solutions. In Manimekalai [13], a review on the various biometric methods used in health care information systems was carried out. The authors also designed a new method for biometric health care information system for state of humans that is unconscious. The authors noted that health care system has entered into the cloud and particularly took the problem of heart attack patients as a case study.

Diaz-Palacios et al. [14] featured a privacy policy to access a central health record database using biometric identification. The study was implemented real time and the results confirmed an improved latency time of 19.8 s with above 200,000 patients record in the system. Mirembe [15] investigated and analyzed a number of telemedicine, e-health, and wellness (TEW) systems. The findings from this study show that majority of the wireless sensor-based technique focuses on engineering-related issues. Esam et al. [16] applied bimodal (face and fingerprint) authentication measures to secure patient medical information within a system. He et al. [17] proposed a novel approach using mutual access and tri-factor authentication protocol for body area networks in health care application. The paper concluded that the proposed protocol is better than the existing bi-factor user authentication schemes. It was proposed in [18] that solutions to secure communication between the patient, its device, and the network channel are implementable. In concluding the related works session, use of biometric technology with heart sounds was engaged in the technology of authentication in body area networking system [19].

# 3   System Modeling and Design

The HIS was modeled using Unified Modeling Language (UML) and it consists of activity and class diagrams. The activity diagram in Fig. 1 shows the process flow of users' registration, system access through consultation, drug administration, and patient treatment. First, patient has to register and receptionist checks for doctor's availability. Thereafter, patients consult available doctor and room is allocated to patient if recommended for admission. Continuous medication is given to patient and is discharged if certified healthy by a doctor.

A deployment diagram was also created with four objects, namely from bottom-database server, printer, hospital local server, and desktop client. The desktop client serves as the user interface through which patients, doctors, and administrators interact with the system. The hospital local server provides application services, the different modules that make up the software, and also storage of medical data. The users are expected to place their right thumb finger on the fingerprint scanning machine. Improper placement of finger on the machine will report, "please place enrollment finger on finger print scanner." The class diagram comprises of five objects which are Enroll, Patientinfo, Assigned, Inpatient, and Ptreatment. One non-medical staff can register one patient. A non-medical staff and nurse go through the process of enrollment with the system. Each doctor also gets enrolled with the system. One doctor can admit one patient and can also treat one patient. One nurse can give medication to one patient at a time (See Fig. 2).

The HIS system was implemented using a web-enabled tool-Visual Basic.Net for the user interfaces, and Microsoft Access as the backend. The system has four essential modules, namely [5]:

i.   **Registration**: The registration module takes care of the patients registered in the reception area. It also handles assignment of patients to any doctor on duty.
ii.  **The Nurse**: In this module, the nurse is able to view, give medication and invoice to the patients.
iii. **The Doctor**: With the doctor module, the doctor is able to give medications to patients, bill patients, admit patients, and also able to verify patients.
iv.  **The Admin**: The admin module is responsible for all the modifications that are made in the backend and to enroll users. For each user enrolled in this module, an access privilege selects their status. The other screen shots of the HIS system reported in this paper are contained in Figs. 3, 4, 5, 6, 7, and 8.

Figure 3 shows the home page of the HIS. The page contains patient, doctor, tools, and HMS. Figure 4 contains the enrollment environment for the user. The biodata captured in addition to the fingerprint includes title, last name, first name, and job status. The user has the option to click on enroll user button or cancel entering.

Figure 5 performs the role of assigning patient to doctor. The verify patient's screen (see Fig. 6) is used by the doctor to verify patients, and ensure that patient is who he/she claims to be.

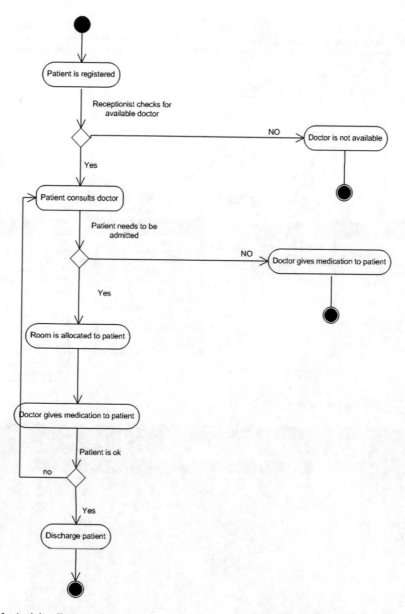

**Fig. 1** Activity diagram of the system

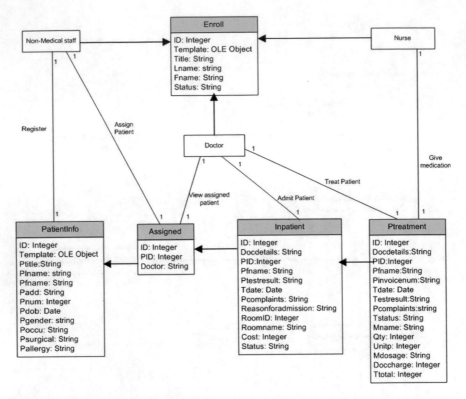

**Fig. 2** Class diagram for the system

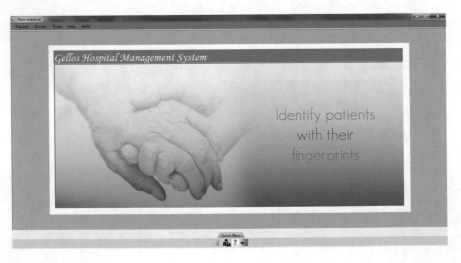

**Fig. 3** HIS home page

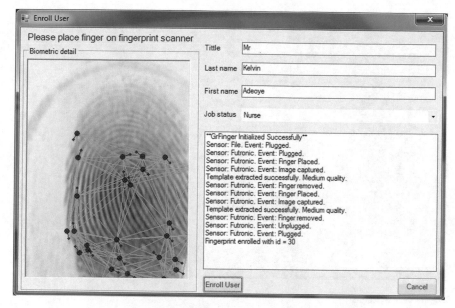

**Fig. 4** Enroll user form

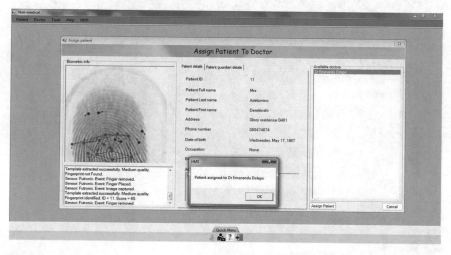

**Fig. 5** Assign patient to doctor form

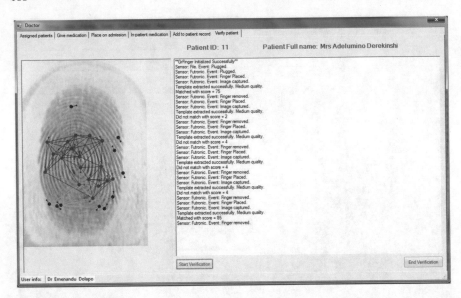

**Fig. 6** Verify patient

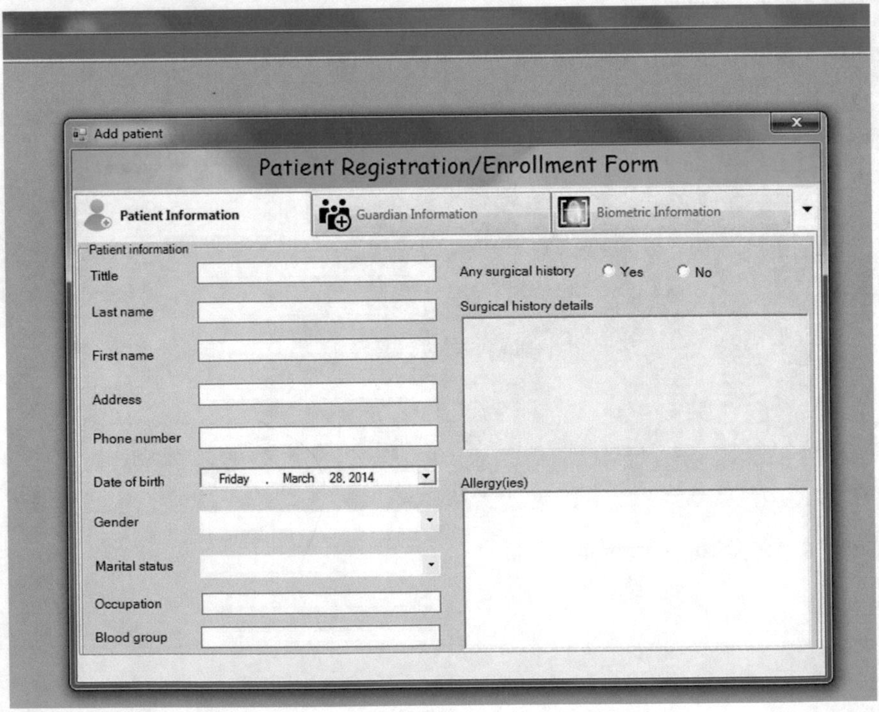

**Fig. 7** Registered patient form

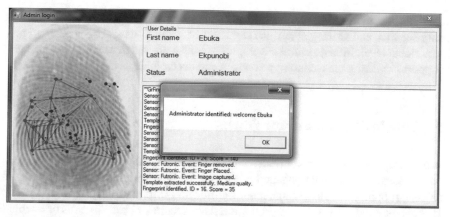

**Fig. 8** Admin identified form

The non-medical staff uses the interface in Fig. 7 to register and enroll users into the HIS system. The interface contains title, first name, last name, gender, date of birth, address, phone number, marital status, occupation, and blood group. The other data requested for patient registration include surgical history details and allergies. The screen in Fig. 8 is the admin screen and is utilized by the doctor to select administrators (after authentication). This takes place before the system grants the admin access to users enroll. For every successful enrollment, the message administration identified welcome users' first name is displayed as a pop-up screen.

## 4 Discussions and Recommendations

The implementation of security in the health care sector has improved safety rules for all the health practitioners including the patients [20]. However, there are challenges that may reoccur, such as the use of passwords for protection against unauthorized users can in addition lead to false sense of security. Applying simple passwords can allow entry of unauthorized user into the system. Patient information are vital keys to administering proper care; thus, an incomplete or wrong patient health records can lead to wrong medication [3].

The use of biometric technology in health applications has great benefits over other means of authentication methods. Password and pin can easily be forgotten and is subjected to theft. In developing countries including Nigeria, the use of biometrics as a means of authentication in e-health sector is still at its low ebb. This article is intended to serve as a security model for protecting the integrity of users of e-health data.

Considering the level of security challenges in the society, it has become pertinent to make some recommendations to government of nations and industry stakeholders in the security sectors, particularly when it comes to authentication of users into accessing sensitive human information. Government can play a role in providing an enabling environment and policies in form of legislation guiding the conduct of data security and sharing at not just the level of institutions and agencies, but also at national level. Infrastructural development is also an area that desires government support in making the technology services and accessibility available to all.

## 5   Conclusion and Future Works

In the computer science, this current study falls under the category of implementation-based research. The study engaged a research method in computer science known as design-build-evaluate to implement the hospital information system (HIS). The HIS can be used by health care professionals to secure and keep records of hospital patient. The expectation of the system is to enhance the efficiency and effectiveness of hospital information management systems. The users' confidence of the system is enhanced following the integration of biometrics into HIS. The future works on this biometrics research in e-health system would comprise one or more of the following: voice-based biometrics in e-health, DNA analysis, neural wave analysis, and skin luminescence [21]. A secured mobile-based biometrics for the able-bodied patient and voice biometrics for the visually impaired patient would also form part of the future works.

## References

1. Mesmoudi, S., Feham, M.: BSK-WBSN: biometric symmetric keys to secure wireless body sensors networks. Int. J. Netw. Secur. Appl. (IJNSA) 3(5), 155–166 (2011)
2. Ikhu-Omoregbe, N.A., Azeta, A.A.: A voice-based mobile prescription application for health-care services (VBMOPA). Int. J. Electr. Comput. Sci. IJECS 10(02), 73–78 (2010). Retrieved 12 Apr 2010 from http://www.ijens.org/1010302-5454%20IJECS-IJENS.pdf
3. Mogli, G.D.: Role of biometrics in healthcare privacy and security management system. Sri Lanka J. Bio-Med. Inform. 2(4), 156–165 (2011)
4. Abayomi-Alli, A., Ikuomola, A., Aliyu, O., Abayomi-Alli, O.: Development of a mobile remote health monitoring system–MRHMS. Afr. J. Comput. ICT 14–22 (2014)
5. Azeta, A.A., Iboroma, D.A., Azeta, V.I., Igbekele, E.O., Fatinikun, D.O., Ekpunobi, E.: Implementing a medical record system with biometrics authentication in E-health. In: 2017 IEEE AFRICON, Cape Town, South Africa, 18–20 Sept 2017 (2017)
6. Ikhu-Omorebe, N.A., Azeta, A.A.: Design and deployment of mobile-based medical alert system (Chap. 10). In: Handbook on E-Health Systems and Wireless Communications: Current and Future Challenges. A Book published in the United States of America by Information Science Reference (an imprint of IGI Global), pp. 210–219. IGI Global (2012)
7. Bazin, A.: Biometrics for patient identification—a US case study. ID World Abu Dhabi, 18–19 Mar. HealthTech Innovation, Fujitsu (2012)
8. Jhaveri, H., Sanghavi, D.: Biometric security system and its applications in healthcare. Int. J. Tech. Res. Appl. (2014)
9. Wang, S., Liu, J.: Biometrics on mobile phone. Recent Appl. Biom. (2011). Retrieved from www.intechopen.com

10. Cheng, X.R., Li, M.X.: The authentication of the grid monitoring system for wireless sensor networks. Prz Elektrotechniczn (2013)
11. Darrell, S.: Biometrics—implementing into the healthcare industry increases the security for the doctors, nurses, and patients. Thesis for Masters Degree Information Assurance (2013)
12. Zuowen, T.: An efficient biometrics-based authentication scheme for telecare medicine information systems. Przegląd Elektrotechniczny 200–204 (2013)
13. Manimekalai, S.: Study on biometric for single sign on health care security. Int. J. Comput. Sci. Mob. Comput. 3(6), 79–87 (2014)
14. Diaz-Palacios, J.R., Romo-Aledo, V.J., Chinaei, A.H.: Biometric access control for e-Health records in pre-hospital care. EDBT/ICDT, 18–22 Mar, Genoa, Italy (2013)
15. Mirembe, D.P.: Design of a secure framework for the implementation of telemedicine, e-Health, and wellness services. Masters thesis delivered to Radboud University Nijmegen Security of Systems (2006)
16. Esam, O.A., Ngwira, S.M., Zuva, T.: Biometric authentication system to protect sensitive medical data. Bimodal Biometrics for Health Care Infrastructure Security. In: Proceedings of the International Multi Conference of Engineers and Computer Scientists, vol. I, IMECS, 12–14 Mar 2014, Hong Kong (2014)
17. He, C., Bao, S., Li, Y.: A novel tri-factor mutual authentication with biometrics for wireless body sensor networks in healthcare applications. Int. J. Smart Sens. Intell. Syst. 6(3), 910–931 (2013)
18. Ivanov, V.I., Yu, P.L., Baras, J.: Securing the communication of medical information using local biometric authentication and commercial wireless links. In: Proceedings of the 14th International Symposium for Health Information Management Research at Kalmar in Oct 2009. Health Inform. J. 16(3), 212–223 (2009)
19. Andreeva, E.: Alternative biometric as method of information security of healthcare systems. In: Proceeding of the 12th Conference of FRUCT Association Department of Information Security Technologies, pp. 210–214 (2012)
20. Shawl, D.: Biometrics—implementing into the healthcare industry increases the security For the doctors, nurses, and patients. Thesis for Masters Degree Information Assurance (2013)
21. Duquenoy, P., George, C., Kimppa, K.: Ethical, legal and social issues in medical informatics (Chap. 11). In: Biometrics, Human Body, and Medicine: A Controversial History, vol. 2, issue 6, pp. 15–20. IGI Global (2008). 8163, www.ijtra.com

Printed in the United States
By Bookmasters